DANIELLE CLODE

A FUTURE IN FLAMES

untapped

ABOUT *UNTAPPED*

Most Australian books ever written have fallen out of print and become unavailable for purchase or loan from libraries. This includes important local and national histories, biographies and memoirs, beloved children's titles, and even winners of glittering literary prizes such as the Miles Franklin Literary Award.

Supported by funding from state and territory libraries, philanthropists and the Australian Research Council, *Untapped* is identifying Australia's culturally important lost books, digitising them, and promoting them to new generations of readers. As well as providing access to lost books and a new source of revenue for their writers, the *Untapped* collaboration is supporting new research into the economic value of authors' reversion rights and book promotion by libraries, and the relationship between library lending and digital book sales. The results will feed into public policy discussions about how we can better support Australian authors, readers and culture.

See untapped.org.au for more information, including a full list of project partners and rediscovered books.

Readers are reminded that these books are products of their time. Some may contain language or reflect views that might now be found offensive or inappropriate.

In memory of

William Marsden
Peter Singleton
Maurice Atkinson
Stuart Duff
Neville Jeffery

Panton Hill Fire Brigade
16 February 1983

CONTENTS

I · The Fiery Continent

1 · The Burning Bush · **3**
2 · From Ashes to Ashes · **11**
3 · Fire-Stick Farming · **27**
4 · A Walk in the Park · **47**

II · The Human Impact

5 · A Growing Risk · **65**
6 · Fire in the Outback · **79**
7 · Living in the Forest · **99**
8 · Urban Impact · **109**
9 · The Big Ones · **123**

III · The Fire Factors

10 · Ignition · **143**
11 · Fuel · **153**
12 · The Bush Brigades · **177**
13 · Be Prepared · **191**

IV · Response

14 · Near Miss · **227**
15 · Facing the Flames · **231**
16 · Leaving Late · **249**
17 · Shelter · **267**
18 · The Need to Know · **283**

V · The Future

19 · Recovery and Resilience · **299**
20 · What Next? · **311**

Acknowledgements · **325**
Notes · **328**
Copyright · **346**

I · THE FIERY CONTINENT

1 · THE BURNING BUSH

My home is in a forest's heart. Every morning I float into consciousness in a sea of grey-green that stretches down the gully below my bedroom window. Cocooned from the outside world, my home in the forest on the northern outskirts of Melbourne is a sanctuary of quiet contemplation.

But my home does not always feel so safe and secure. Every summer, hot and windy days bring a sense of foreboding and anxiety. The dried and yellowed grass crackles underfoot and the ground is littered with the discarded twigs, leaves and bark of the eucalypts. The trees perspire clouds of flammable perfume in the heat.

The view from my bedroom windows may be a picture of lush green serenity but the thin vertical tree trunks betray a violent history. Grey stringybarks and box trees dominate the landscape, wrapped tightly in their shaggy, unkempt bark. An occasional stripe of black denotes the trunk of a wattle topped with delicate feathery foliage. Towards the gully, smooth white candlebarks and manna gums carelessly shed their outer vestments onto the ground beneath them, lifting their naked satin limbs skyward.

Most of the trees are relatively young, still thin and spindly, crammed close together. But here and there are older survivors, with their thickset trunks and broad, arching canopies. If you look closely you can see black scars running through the bark, on the edges of fissures and cracks, the telltale signs of fires, large or small, that have burnt this bush before.

Today I can see only trees from my house but in 1962 the

scene would have been very different. Looking south into what is now a wall of vegetation, I could have seen the burnt remnants of Panton Hill, 8 kilometres away across a blackened hillside. One resident, Marjorie Smith Motschall, who lived nearby, described the scene then:

> The devastation was unbelievable. Every acre of pasture, every chain of fencing, burnt. Miles of blackened trees standing in ashy bare ground where green trees had been. At Butterman's Track the ground showed all the scars and pits of the Caledonian mining era. A depressing and hopeless scene. The soft rain after the fire brought a thin coat of green over the burned land, but further hot winds scorched it off again. The humus beneath the trees had been burnt from the soil and will take a long time to recover. There had been no further rain. The people who were living in the 'Black Forest', though their homes were saved, faced a miserable time. Not a bird or animal, all was silent, nothing but charred stumps and logs.[1]

The 1962 fire began innocuously enough when someone tried to smoke out some rabbits in Christmas Hills one Sunday in January. But the surrounding countryside was a tinder box after a dry winter, no spring rain and a hot summer. The local fire brigades, and later volunteers from the city, were called out as the south-easterly winds drove the fire north into the Kinglake National Park and across to the tiny township of Strathewen. Despite their efforts, when the wind changed to a blustery, hot northerly early on Tuesday morning, the fire roared back south along the main road, consuming the towns of St Andrews, Panton Hill and Watsons Creek, skirting around Kangaroo Ground before finally running into a cool change, and welcome rain, on the outskirts of Warrandyte. This fire burnt 180 homes, thousands of hectares of land and claimed two lives.

Devastating though it was for the people who lived through it, this fire was just one of many that raged across the Australian countryside that summer. In a week, Victoria lost 400 homes and 14 lives to fires. Historically 1962 was just one of many 'extreme' seasons that periodically afflict Australia. Every year, more than 500 bushfires start in Victoria alone. In a severe season, such fires each burn an average of a thousand hectares, resulting in around half a million hectares of scorched earth. Even in a mild season, an average of 18 000 hectares are burnt each year in Victoria. Across Australia some 270 million hectares, or 40 per cent of the land mass, are prone to bushfire damage. Essentially, all of Victoria, Tasmania, the ACT and most of New South Wales and Queensland are classified as fire prone. Smaller areas of South Australia, the Northern Territory and Western Australia are vulnerable only because of their greater expanses of desert.[2]

According to Emergency Management Australia's disaster database, 815 people have died in bushfires in Australia since European settlement.[3] Despite dramatic improvements in our ability to fight fires, the death toll has remained almost unchanged in that time. On average there are five to ten deaths from bushfires every year. And every decade or two there is a horror year, where the death toll spikes to some new record. These horror years are seared into our collective memories, the days named for their ferocity and tragedy. Twelve people died on Black Thursday in the foothills around Melbourne in 1851. Another twelve people died on Red Tuesday in the Victorian region of Gippsland in 1898. Forty people died on Black Sunday in 1926 when fires swept through the mountain forests of eastern Victoria. Seventy-six people died on Black

Friday in 1939 when these same forests were again swept with fire. Forty-eight people died in fires that razed central Victoria in 1944, burning over a million hectares of grassland and 160 hectares of forest. When fires encroached on the outer suburbs of Hobart on Black Tuesday in 1967, 61 people perished. And on Ash Wednesday in 1983, 75 people died across the states of Victoria and South Australia, many on the outer fringes of the cities of Melbourne and Adelaide.

Things are not getting any better. In 2009, the worst fires on record savaged those same Victorian mountain forests that have been the site of so many tragedies in the past along with areas of Gippsland and central Victoria. In the inferno that roared through the hills with an unprecedented speed and ferocity, 176 people died. Everyone had hoped that a tragedy on the scale of Ash Wednesday could not happen again. We had had bad fires since then, in Canberra, Sydney and on South Australia's Eyre Peninsula. Lives had been lost, but not on the mass scale of the past. Many hoped, imagined, perhaps wishfully, that these mass deaths would not happen again.

Black Saturday proved that all wrong.

We are running out of days in the week to commemorate these disasters.

Over the past two centuries, more than a million Australians have been directly affected by fire. Thousands of houses, farms and properties have been destroyed by bushfires.[4] The cost of each major fire runs into the millions. The personal devastation and trauma is immeasurable.

Every year our firefighters get better and better at fighting fires. The fire authorities buy bigger helicopters, they purchase newer trucks, they train more firefighters, they improve communications, they respond faster and more effectively than ever before. Every year, there are media campaigns, street and community meetings, pamphlets and DVDs to warn people about the dangers of fires. There are building codes and vegetation management overlays that are designed to make sure that properties built in high-risk areas are kept safe, are easier to defend and are better able to protect their inhabitants.

But every year there are more and more people living in high-risk areas. Per capita, fewer people are dying in bushfires but the total figure remains the same. Every year there are more and more fires being lit. Every year seems to be getting drier and drier. As fast as we improve our capacity to fight fires, we seem to increase our vulnerability to them.

The fact is that we are getting better at containing and controlling small fires. No longer do we rely on rain to put fires out. Very often firefighters can actually extinguish fires, even moderately sized fires, something that was unheard of in the past. But severe fires are beyond the ability of anyone to control. There is no question of stopping fires like those on Black Saturday once they get going. It becomes a question of surviving them.

In a severe fire, there is no fire brigade to rescue you. There are not enough trucks or helicopters to protect every home and they couldn't get to you even if there were. In a severe fire, the first person to know there is a fire in your area is probably going to be you. Big fires can move so fast and so unexpectedly that by the time authorities know where the fire is, it has already raced far ahead. There is no white knight on a big red truck. You have to face the dragon yourself.

Every summer, fire authorities across the country run campaigns to help people prepare for the risk of fires. Enthusiastic presenters explain in patient detail, the importance of having a plan—several plans—what it means to leave early, what preparing to stay and defend really means. Preparing for fire is not easy; it is complicated and unpleasant to think about. It is inconvenient and often expensive. It involves making decisions we'd rather not have to. It involves making commitments for an event that may never happen.

Every summer we see fires burning on the news. We see footage of people waving garden hoses at burning fences, running through smoke in singlets and bare feet. Children and pets are rushed to cars as flames lick at their backyards. We see people standing on their roof pointing at the approaching fire. They clearly didn't have an appropriate plan, they didn't leave early enough and they are obviously not prepared.

After the 1939 fires, Judge Leonard Stretton concluded his investigation with the observation that the victims 'had not lived long enough'.[5] He wasn't commenting on their youth or their lives cut tragically short. He was commenting on the Australian community's failure to recognise the risks our native vegetation, climate and weather place us in. We had not lived long enough to learn how to prepare, what was safe and where was not safe. In truth, he was probably being a bit generous. By 1939, Victorians in particular and Australians in general were already well versed in the frequency and severity of fires. How many centuries do you need?

Seventy years later, in the aftermath of Black Saturday fires, environmental historian Tom Griffiths returned to Stretton's theme.

> There is a perennial question in human affairs that is given real edge and urgency by fire: do we learn from history? Testimony from the 1939 and 2009 fires suggests that there is one thing that we never seem to learn from history. That is, that nature can overwhelm culture. That some of the fires that roar out of the Australian bush are unstoppable. As one fire manager puts it, 'there are times when you have to step out

> of the way and acknowledge that nature has got the steering wheel at the moment.' It seems to go against the grain of our humanity to admit that fact, no matter how severe are the lessons of history.[6]

Two centuries is surely long enough to learn the lessons of our environment. And yet we have not. We are still unprepared. We still live in high-risk areas, in houses that cannot protect us and even place us in danger. We still don't leave early enough. We still expect the white knight on a red truck to save us, or at least to tell us what to do. We still think we can 'wait and see'—and leave if the fire gets too bad.

It is the reality of natural disaster like fires that sometimes no amount of preparation is enough. Even the most prepared person in the world worries that their preparations may not be adequate, that the 'big one' will just be too big. That something completely unexpected, some random accident, will intervene and the best laid plans will go astray. Sometimes shit happens.

But these are the exceptions. Over the years we have learnt what is dangerous, what activities place us at the highest risk. We know how dangerous it is to be caught in the open during a fire, how risky cars are, how you need to actively defend your home while sheltering inside. In most fires, deaths occur while people are evacuating late, caught in the open or passively sheltering. The better your preparation, the better your planning, the better your chances of survival. There are no guarantees, but the odds certainly improve dramatically.

So why, after so many years of experience, are people still dying in bushfires in Australia? Is it just that too many people aren't prepared? If so, why don't they prepare? What does it take to get us to take fires seriously?

Or are our preparations simply not enough? Are fires

getting more dangerous? Is climate change shifting the goal posts? Is it possible to live with fire?

As I sit on the verandah of my home in the forest, I worry about my own preparation and my own decisions. Am I foolish to expect to be able to defend my house, with its fire pumps and defendable space, on my own with two children, while my husband is out on a truck, saving someone else's home? Or am I just placing all our lives in danger? But how would I feel if my house burnt down because of an ember or a spot fire, knowing that I could easily have saved it if I had been there?

I really don't know the answers to these questions. This book is my attempt to find out.

2 · FROM ASHES TO ASHES

IN THE BEGINNING

For millennia the lands had shifted and fused, separating and reconnecting across the changing face of the planet. The continents drifted on incremental voyages as life began its own slow journey from microbial origins to complex organisms. Plants grew and spread over the land, locking up carbon and releasing oxygen, creating fuel and atmosphere, the ingredients for a new chemical reaction—fire.

Beasts with backbones swam, then walked. Gondwana split from its northern neighbours. The world was warm and wet. Lush tropical forests grew and fell into rich decay in the earth, carbonising, forming coal. The continent moved south—fragmenting and splintering—south into the Antarctic Circle. Dinosaurs stalked the winter darkness.

The narrow waters that separated Australia from Antarctica spread wider until the currents that churned around the continental land masses channelled their force through this new short cut. No longer warmed by a tropical deviation, the water grew colder, locking Antarctica in an icy embrace. The temperature dropped, the seas rose, and the world changed.

Australia and New Guinea drifted on alone to the north. A vast shallow sea flooded the southern inland of the continent. The great forests of conifers and cycads declined. Flowering plants overshadowed their ancient relatives. Casuarinas, proteas and beeches flourished. The open forests, adept at catching the long, low rays of the polar sun, closed their canopy as the sun rose above them and the rains fell.

A few volcanoes pockmarked the edges of the continent, but otherwise the ancient soils went unrefreshed, slowly leaching their meagre nutrients. From this harsh bed sprang tough-skinned plants that spurned the phosphorus and nitrogen beloved of soft, leafy greens. Grey-foliaged banksias flowered, leaving woody seed pods for the next generation. Climates changed and the inland sea dried up. The north grew wetter and warmer and the south grew drier and colder. The vegetation diversified.

Lightning cracked across the landscape, just as it had for countless eons but now, in the dry forests, it found fuel. Australia ignited. Species shrivelled under the onslaught of the flames. Others regenerated, stronger and more abundant than before. The eucalypts flourished and spread. Giant beasts looked out over expanding grasslands from their forest sanctuaries and bellowed their disquiet. Tasmania and New Guinea became islands. Humans walked across the land, trailing fire sticks, and the forests and grasslands burnt more often. The soft, green rainforests retreated to the moist gullies and wet mountains. Hard-leaved, thick-barked plants sprouted from the underground tubers they were saving for drought. Others cast their seeds into the rich ashbed and regenerated, phoenix-like, from the flames. The bush became inured to fire. The survivors promoted flames, shedding great drifts of bark onto the forest floor, as if inviting disaster.

The animals also adapted with their changing homes. The great beasts disappeared. Smaller grass-eaters prevailed. Kookaburras flew into the newly burnt forests—cheerful guests for a neighbourhood barbecue. Reptiles and birds, mammals and humans all moved within the mosaic patchwork of habitats created by fire.

Fire became a necessity. Modern Australia was born.

THE LAKES

Lake George, some 30 kilometres north of Canberra, is not so much a lake as a grassy plain: a wide, flat basin rimmed with low, rounded hills. In times of severe drought, the lake dries up completely. At other times it has had enough water to harbour a thriving Murray cod fishery, even its own trawler. For all its inauspicious appearance, Lake George is a significant environmental indicator. It last dried up completely in the Federation drought at the beginning of the 1900s. Today it is dry again.

Beneath the surface of Lake George lies an even more important environmental record. The shallow lake has ebbed and flowed for millions of years, leaving a record of past climates layered in its edges. Within these layers are smudges of charcoal, indicating fires that have spread across the dry lake edges and surface. Over the last million years, the layers of charcoal steadily increase in frequency, as the climate has dried. But around 120 000 years ago, 8.5 metres below the lake's surface, the abundance of charcoal increases dramatically. Fires started burning far more often across the lake's dry bed than they had previously.[1] Something had happened to increase the incidence of fire.

Charcoal records along the east coast of Australia reveal similar patterns.[2] Fire has been evident in the geological record for 470 million years, ever since plants first provided the fuel and the oxygen-rich atmosphere needed for combustion to occur. Initially, the frequency of fire was linked to the amount of oxygen in the atmosphere, and later to the prevalence of grasses in a drier climate.

In recent times, the frequency of fires has increased dramatically. This increase seems to date in some places from as long as 150 000 years ago, and in others to as few as 26 000 years ago.

The same period of time saw a dramatic change in the plants and animals that inhabited the Australian continent. Pollen records trace the retreat of the wet rainforests from many areas, and a corresponding expansion of grasslands and open woodlands dominated by eucalypts, acacias and casuarinas. Numerous studies have analysed, dated and interpreted this changing vegetation and its apparent association with both a drier climate and more fires.[3]

The fossil record also documents the extinction of the last of Australia's megafauna. Between 40 000 and 60 000 years ago, the last of the giant 2-tonne diprotodons, marsupial lions and flightless birds bigger than emus disappeared from the Australian landscape.[4] Something significant must have happened around this time to reshape the landscape and its fauna so dramatically. Was climate change enough to account for both the increasing fires and the changing vegetation? Could increasing fires alone have changed the vegetation and the animals that depended on it? Or was there something else driving these dramatic changes?

Along with all these changes, a new species had appeared on the Australian landscape—one that brought with it an unprecedented ability to manipulate and alter its environment to suit itself. Humans.

FIRE ECOLOGY: A GLOBAL FEATURE, A LOCAL PHENOMENON

Fire is an integral feature of many natural ecosystems all over the world. The South African fynbos heathlands of the Western Cape region are famous for the proteas whose large-headed flowers retain their seeds for months, awaiting regeneration by fire. The long coastal strip of matorral shrubland in Chile is primarily burnt by human activity, yet most of its plants seem fire tolerant and able to resprout after fire. By contrast, the dense, drought-adapted chaparral scrub of California appears to have evolved in response to lightning strikes and is home to large numbers of fire-dependent specialists that release masses of seed after fire. The stony landscape of the garrigue of southern France infuses the region with the grey foliage and aromatic oils of broom, juniper, wild thyme, rosemary, cistus and lavenders—characteristics of drought and fire-adapted vegetation. Particular fire regimes are important to all these ecosystems for maintaining the diversity of plants.

But only in Australia have so many different ecosystems adapted to fire—from tall forests to open woodlands, from heath to grasslands. Strangely enough, it is water that has created Australia's peculiar relationship with fire. Water allows plants to grow, providing the fuel for fire; the absence of water enables fires to burn that fuel. So, despite being the driest continent, it is not the driest parts of Australia that burn the most often, but the areas that are wet part of the year and dry in others. Ironically, the only large areas of Australia rarely affected by fire are the deserts stretching inland from the west coast.

More than 80 per cent of Australia experiences three months of the year without significant rainfall. As a result, Australia is the only continent to have so many ecological communities burnt with varying regularity over a large area and over a long period of time. Only in Australia has fire had such a major effect on so much of the vegetation. Fire has moulded the Australian landscape over thousands of years, and it continues to mould all who choose to live here—plants, animals and humans alike.

THE FIRE LOVERS

Humans love fire. Many animals follow fires, keen to scavenge and prey on the fire's victims or exploit the cleared spaces. No doubt humans did the same in their early history. Unlike other animals, humans commandeered this useful natural phenomenon, taking it home with them to warm themselves, cook their meat and deter predators. Deep in the Swartkrans Cave of South Africa, just north of Johannesburg, the bones of warthogs, antelope and baboons reveal evidence of deliberate cooking by early humans in a hot camp-fire (rather than a cooler bushfire).[5] These ancient remains, 1.5 million years old, suggest that humans have long appropriated fire for their own purposes, once it had been lit by lightning or other natural causes.

Having discovered how to use fire, humans don't seem to have taken long to learn how to create it themselves. Not content with having to tend fires through wind and rain or keep fire sticks smouldering, prehistoric humans soon learnt to use flints to strike sparks into dry tinder. Archaeological remains of human settlements on the edge of the Dead Sea reveal concentrations of burnt flints that seem to have been from ancient hearths or camp-fires 750 000 years old.[6] Having acquired the useful skill of starting fires, human communities are likely to have spread this knowledge rapidly as they radiated out of Africa and across Asia.

Deep piles of ash, 500 000 years old, have been found in the home of Peking Man in the Zhoukoudian Caves of China. Native Americans used fire, not just for domestic cooking and warmth but also to drive game across the plains. Whether deliberately or accidentally, humans appear to have increased the incidence of wildfires wherever they live. It would be surprising if the same pattern did not occur in Australia.

The oldest known human remains in Australia are 40 000

years old. Some settlement sites in the Northern Territory are thought to be 50 000 to 60 000 years old.[7] Because the association between humans and fire is so strong, some researchers argue that increased fire frequency itself is evidence of human presence. The great abundance of charcoal deposits off the Queensland coast, from a time when sea levels were lower, has been used to suggest that Aboriginal settlement may have occurred as early as 140 000 years ago.[8] This begs the question: is the association between increased fires and Aboriginal habitation causal or circumstantial?[9] We must first be sure that Aboriginal people did increase the incidence of fire in the Australian landscape before we can use an increased incidence of fire as evidence of their presence.

FIRE IN THE RAINFOREST

The tropical Atherton Tablelands in the heart of north Queensland are a surprising setting for evidence of declining rainforests and a drying climate. A 200 000-year-long record of pollen and charcoal deposits at Lynchs Crater reveals a cyclical rise and fall of tough, oily-leaved sclerophyllous plants like eucalypts and acacias in relation to rainforest plants. As the climate cools and dries and the sea levels fall, the proportion of eucalypts increases. When the climate is warmer and wetter and the sea levels rise, rainforest species predominate. For much of the record, the charcoal deposits remain much the same. But around 26 000 years ago, burning increases dramatically, perhaps as a result of human activity. It seems that this had the effect of forcing the reduction of rainforest species even more than in previous cycles,[10] providing potent evidence of the potential impact of burning, presumably of human origin, on local vegetation.

Lynchs Crater today is surrounded by rainforest. Left to its own devices, the open farmlands of the Atherton Tablelands would rapidly revert to luxuriant dense rainforests, not to open, fire-prone, eucalypt woodlands. Anyone who lives in this area would be able to tell you why—it is wet. With one of the highest rainfalls in Australia, there is little chance of eucalypt forests predominating over rainforest here. For the last 10 000 years, the Earth has entered what is known as the Holocene interglacial, a period of warmer and generally wetter weather. In response to this climate change, and despite continued burning, the rainforest has returned.[11]

Rainforest rules

Along the north-east Australian coast the rainfall increases, as easterly winds drive moisture-bearing air from the Pacific up the mountains, condensing their burden into heavy rain on the coast and slopes of northern Queensland. The abundance of water suppresses fires, promoting the growth of soft, leafy rainforest plants with little ability to survive either drought or flame.

Even in these warm, wet tropical rainforests, fire sometimes strikes. Lightning might ignite an old dead tree, its aged timbers providing dry fuel. More commonly, fires skirt the edges of the rainforest, making incursions into the drier, tougher, fringe species: the crowded cluster of vines and climbers that seal off the rainforest from the outside world. Such incursions encourage the growth of eucalypts and other fire-tolerant plants. Over time, the rainforest plants reassert themselves, shading out their competitors and re-establishing their own forest microclimate. Without fire, the rainforest plants are all conquering, enveloping the damp landscape in a thick, green, impenetrable cloak.

Along the coastal and inland plains, fire provides a brief respite from the dominance of the rainforests. Pockets of grassland around Eubenangee Swamp and Princess Charlotte Bay depend upon hot fires to halt the invasion of melaleuca forests dominated by the broad-leaved tea-tree. This vigorous plant suckers readily, rapidly creating an impenetrable thicket of shrubbery that shades out the grasses. Frequent fires burn off the young suckers, preventing the dominance of the tea-trees, while the grass reshoots from basal nodes. But, if the tea-tree suckers avoid fire long enough to grow over a metre high, they can survive the cool grass fires and only a hot, severe fire is able to kill them. Such hot fires, capable of reaching into the canopy, are rare in the humid tropics, even in the dry season.

CHARCOAL IN THE MOUNTAINS

A study of charcoal sediments from Goochs Swamp in the Blue Mountains also supports the idea that climate is the primary driver of vegetation change, and also fire frequency. This research found that fluctuations in fire frequency over the last 14 000 years can best be explained by changing climate.[12] In times when the climate fluctuated between drought and wetter years (what we now know as the Southern Oscillation, or El Niño cycle), fire frequency was also highly variable. When the climate was stable, and probably moister, fire frequency declined. These patterns match similar ones found in South-East Asia.

The work at Goochs Swamp found a dramatic increase in charcoal loads, associated with frequent and intense fires, from 5000 to 3000 years ago. This is unlikely to have been related to human activity, since a similar increase is found in New Zealand sediments in areas where no-one lived. This time frame also coincides with a gap in the deposition of Aboriginal artefacts at many sites in the Blue Mountains, suggesting that people were not living in the area during this time of high fire activity.[13]

The conclusion that climate, rather than humans, has driven both increasing fires and changed vegetation has also been drawn from studies of charcoal sediments at Lachlan River in central New South Wales and at Darwin Crater in south-west Tasmania. Indeed, at Darwin Crater, there is no increase in burning or reduction in rainforest associated with Aboriginal settlement around 35 000 years ago. Nor does charcoal increase in the sediments of Kangaroo Island in South Australia or Hunter Island or Flinders Island in Tasmania during the times when they were inhabited.[14]

CLIMATE CHANGE

The close association between climate and fire seems to rule out human impact in Australia's fire regimes and associated changes in vegetation. But what if human activity can also change the climate? While we are familiar with this concept in the modern industrial era, recent research suggests that even pre-industrial societies may have had a significant impact on the Earth's climate. The global clearing of forests for various cultivated grasslands and the dramatic increase in methane-producing livestock may well have contributed to global warming at a much earlier time than previously thought. Could human burning patterns have changed Australia's climate?

The association between fire, climate and vegetation is highly complex. One of the features of the Australian climate early in human history was the unexpected failure of the monsoon rains to bring water to inland Australia. Lack of rain dried up the inland lakes and was associated with a change from a shrubby, more treed environment to the sparsely grassed desert landscape we are familiar with today. This change appears to have played an important role in the extinction of many megafauna species. Many of the species that became extinct were browsers or depended on forests and woodlands, while those that survived tended to be grazers, favoured by the expanding grasslands.[15]

It is tempting to assume that the failure of the monsoon rains was responsible for the changing vegetation but, remarkably, some researchers argue that it was the vegetation change that halted the monsoon rains.[16] The global climatic conditions that should create the monsoon rains over inland Australia were highly favourable 50 000 years ago. So why did the rains fail?

Rainfall, it seems, is particularly sensitive to vegetation and soil conditions. Vegetation changes the way moisture is recycled, alters the amount of light reflected from the sun and how much the land heats up. Less vegetation would cause greater convection—more hot air rising—over inland Australia that, in turn, would increase the amount of air flowing onto the northern lands from the sea, which would have pushed the monsoonal trough south and away from the inland areas. If changing vegetation was the cause of the dry conditions, rather than the result, what changed the vegetation? These researchers suggest that it was the arrival of humans, whose use of fire changed the shrubby forests to open grasslands.[17]

This controversial argument is supported by Peter Latz, botanist and long-time resident of arid Australia. Latz argues that the encroaching deserts and increasing aridity of inland Australia has been brought about by the use of fire by Aboriginal people over the last 60 000 years. Latz noticed that, compared to other continents, Australia's deserts have relatively few species and those that live there, while drought tolerant, are not the true desert specialists like the American cactuses or the African aloes and agaves. Many Australian desert species are also found on the coast, suggesting that they are generalist invaders of a new habitat rather than specialist residents of a long-term environment. Latz argues that high levels of hybridisation between species and erratic patterns of distribution also suggest recent incursions by plant species, rather than an ecologically stable community.[18]

Grass and fire

Grasslands are important ecosystems for many species, particularly humans, but they are not the elixir of life in all circumstances. With age, grass rapidly becomes a tough, prickly inedible mass. The most drought-tolerant grasses, common across much of the desert country in inland Australia, are collectively known as spinifex grasses. Nothing much eats old spinifex. Spinifex grasses don't even begin to break down until they are about 15 years old—a remarkably long life span for a grass. Without fire, spinifex dominates the grasslands, turning them into an inhospitable monoculture. Fire, particularly in patches, promotes diversity, killing off some species and allowing others to grow—maintaining an even playing field. Fire gives the kangaroo apples a chance to grow, the nicotiana bushes the opportunity to germinate. Even the spinifex reshoots in a lush profusion of new growth. This basic mosaic of plant diversity supports a profusion of animal species.

Spinifex grasses—primarily the *Triodia* tussock grasses—cover almost half of Western Australia and most of the Northern Territory and South Australia. The true spinifex species are only found along the coast. Inland, through the arid regions, it is *Triodia* that dominates.

In the north, monsoonal rains promote lush growth that then cures yellow and dry in the winter. Spear grass, which grows 4–5 metres high, is a potent fire risk in northern regions. This growth, combined with dry thunderstorms that crackle lightning across the northern skies, makes the north-west the most fire-prone corner of a fire-prone continent. Fly over this land of an evening and you will see a tracery of fine glowing lines, slowly singeing their way across the land. In the daytime, great plumes of smoke billow up into the atmosphere, their baggage of fine particles seeding the thunderheads to create the lightning that sparks the fires. Fire creates fire from cloud.

The fires burn quick and cool here. The dry northern winter is cooler and more humid than the summers of the south. There is less fuel. Not only are the grasslands less densely vegetated than the forests but they also burn regularly, even annually, reducing the fuel load they carry. Fires in the Northern Territory are often vast in scale and frequent, but they are nowhere near as fierce as the forest fires of the south.

HUMANS OR CLIMATE?

Palaeontology suggests two plausible options for the relationship between humans and fire in Australia. People may have increased the incidence of fire, changing the vegetation, which dried the climate, both circumstances that would favour further fires. Alternatively, the arrival of people may have simply coincided with a changing climate that dried the continent and altered the vegetation, which in turn increased the incidence of fires.

Palaeontology, by its nature, provides only a crude measure of change. It measures the end product of change, not the process by which it was produced. The way in which charcoal is deposited in the fossil record is influenced by a wide range of factors, including wind, rainfall and erosion; the manner in which the charcoal records are interpreted is also open to a great deal of debate.[19] For example, one theory is that burning by humans in Australia increased erosion,[20] which would not only change the vegetation but potentially also the way in which the evidence of fire was preserved.

Even if we assume the charcoal records are comparable and accurate, the charcoal remains of a bushfire lit by lightning are the same as the remains of a bushfire deliberately lit by people. Even if there was no change in the overall rates of burning due to the presence of humans, it does not mean that Aboriginal people did not start fires or use fires systematically. Lightning may not start fires as often as humans but it remains a significant cause of ignition across the continent. Land that is already burnt will not burn again until it has regenerated. It is possible that climate dictates the overall level of burning, with the balance of human to natural ignitions sharing out a climatically determined 'quota' between them. What humans don't burn, nature does, but the crude end measure, in terms of the total number of fires or amount of charcoal, remains the same. Aboriginal Australians may or may not have altered the total number of fires but it is certainly probable that they altered the manner and timing of those burns.

Fire is an integral part of the cycle of life and death in the Australian bush. It is so for many of our plants and animals and it seems to have become so for the first human inhabitants.

Irrespective of whether Aboriginal people altered the fire regimes of Australia on their arrival or simply adapted to them, there is no doubt that fire plays and played an important part in traditional cultural life. The advantages of burning would have been obvious, particularly in the wetter parts of Australia, where fire could be introduced to areas where it might not naturally occur. Yibarbuk, a Maningrida man from Arnhem Land, says that traditional people

> burn to hunt, to promote new grass which attracts game, to make country easier to travel through, to clear the country of spiritual pollution after death, to create firebreaks for later in the dry season and a variety of other reasons which 'bring the country alive again'.[21]

The smell of the smoke is said to please the spirits of the ancestors. After death, crows try to chase the spirit away from the spirit land but the hawks and falcons who follow fire defend the spirit and help it into the spirit land, because it was once a living person who burnt the country for them.

For Aboriginal Australians, the sight of smoke provides a sense that the country is being looked after. Smoke means people—a sign that the land is being cared for. Smoke indicates more than just occupation—it is a sign of ownership, residency or custodianship.

Yet for the first Europeans who arrived in Australia, fire indicated only the presence of people, not land ownership. Europeans define land ownership in relation to working the land—ploughing, tilling, grazing, sowing and harvesting. For Europeans, fires belong in the hearth, in the kitchen. They indicate domesticity, not land management. Perhaps if early Europeans in Australia had seen the signs of smoke they frequently reported as evidence that this was an occupied land—a land that was cared for, tended and owned—our history might have been very different.

3 · FIRE-STICK FARMING

Fire, grass, kangaroos, and human inhabitants, seem all dependent on each other for existence in Australia; for any one of these being wanting, the others could no longer continue. Fire is necessary to burn the grass, and form those open forests, in which we find the large forest-kangaroo; the native applies that fire to the grass at certain seasons, in order that a young green crop may subsequently spring up, and so attract and enable him to kill or take the kangaroo with nets. In summer, the burning of long grass also discloses vermin, birds' nests, etc., on which the females and children, who chiefly burn the grass, feed. But for this simple process, the Australian woods had probably contained as thick a jungle as those of New Zealand or America, instead of the open forests in which the white men now find grass for their cattle, to the exclusion of the kangaroo, which is well-known to forsake all those parts of the colony where cattle run. The intrusion therefore of cattle is by itself sufficient to produce the extirpation of the native race, by limiting their means of existence; and this must work such extensive changes in Australia as never entered into the contemplation of the local authorities. The squatters, it is true, have also been obliged to burn the old grass occasionally on their runs; but so little has this been understood by the Imperial Government that an order against the burning of the grass was once sent out, on the representations of a traveller in the south. The omission of the annual periodical burning

> *by natives, of the grass and young saplings, has already produced in the open forest lands nearest to Sydney, thick forests of young trees, where, formerly, a man might gallop without impediment, and see whole miles before him.*
>
> Thomas Mitchell, 1848[1]

STONE AGE POLITICS

The debates surrounding the Aboriginal use of fire are spirited and fierce. Why does it matter so much? Much of the modern debate on fire and our landscape rests on assumptions about how fire affected the landscape before European settlement. What does 'natural' vegetation look like? What is 'natural' for the bush?

Unfortunately, these are not simple questions. Essentially, they rest on the question of whether Aboriginal people used fire, how much they used it, and how much that fire-use changed the landscape. The answers to these questions fall at the intersection of two politically and ideologically charged debates: conservation, and our interpretation of traditional practices.

Some argue that Europeans halted Aboriginal burning practices and caused grasslands to become dense, fire-prone forests. Others argue that early Europeans were themselves 'pyromaniacs' who constantly and repeatedly set fire to the forests. Some people believe that fire is essential to maintain Australia's natural ecological diversity. Others believe fire is just another way that humans disturb and disrupt ecological functioning.

The arguments can be characterised by the two extremes. At one end is the view that Aboriginal people did not increase natural burning and did not significantly alter the landscape or vegetation of Australia. The background to this position is the notion of the indigenous people of the world being, like Rousseau's noble savage, in ecological balance or harmony with nature. But it also implies that the first Australians were passive occupants of the land who did not actively manage the land (at least in relation to fire) and that Aboriginal people cannot be held accountable for fire-induced changes to the landscape or the extinction of the megafauna. The conservation consequences of this line of reasoning also suggest that controlled burning is not a natural part of the Australian ecosystems and should not be part of modern conservation practice.

The alternative extreme is that Aboriginal Australians actively and deliberately used fire to manage and alter the landscape. This scenario depicts the first Australians as active land managers (which aligns more comfortably with Western concepts of property ownership). As a consequence, Aboriginal land managers cannot be seen as necessarily being in balance with nature—rather, they have created a new balance. In creating this new balance, there is the potential to have contributed to species extinction or to have changed water or nutrient cycles. This position can be taken to imply that controlled burning needs to be a part of modern land management practices in order to restore ecosystems to their 'pre-European' functioning or retain it.

If Aboriginal burning practices contributed to the changing climate and vegetation cycles seen over the last 10 000 years, then continuing traditional burning practices may actually be contributing to the increasing desertification of the Australian continent, making what was already a fire-prone continent even worse.

This continuum of views does not always offer comfortable

seats for any of its participants. One person's competent land manager may be another person's environmental vandal, irrespective of whether they are of European or Aboriginal descent. Often our preconceived notions are contradictory—the belief that Aboriginal people did nothing with the land may be held alongside the belief that they were pyromaniacs. Others may hold deep-set beliefs about the nature of a pristine Australian bush that do not conform with the idea of active traditional management.

In order to navigate our way through these perilous waters, we need to look at the historical evidence of Aboriginal burning and its effect on the vegetation. Unlike the broad, sweeping brush strokes of palaeontology and archaeology, historical records give us fine-grained, localised details that hint at broader events and activities.

FIRE IN THE ARCHIVES

The earliest written accounts of fire in the Australian landscape come from the first European explorers and settlers. For them, fire signified the presence of humans. As explorers of new lands, the presence of other humans was an important feature of the landscape, not least because it indicated the availability of fresh water, always an overriding concern for sea captains and inland expeditions alike. Smoke meant people and people meant water.

References to fires are common in the journals of both early coastal explorations and the later inland explorations of the continent. In 1696, the Dutchman Willem de Vlamingh sighted smoke on the mainland, a few kilometres away from his anchorage at Rottnest Island. His crew later noticed smoke rising at many places and when night fell, they could see 'many fires burning along the entire mainland coast'.[2]

Eighty years later in 1770, the Englishman James Cook referred frequently to 'smooks' along the eastern Australian coast. Cook named Smoky Cape after seeing 'a great quantity' of smoke, which might have been a bushfire of some kind. Another 'great deal' of smoke further north near Bundaberg proved to be ten small fires near one another, with cockle-shells nearby.[3] Indigenous campsites often comprise numerous small camp-fires rather than a single large one, allowing individuals to sleep comfortably between the small fires and be warmed both front and back. So it is difficult, from these descriptions, to know whether sighting of many smoke sources along the coast is just an observation of camp-fires, or something else.

CAMP-FIRES AND COOKING

Camp-fires are mentioned by many early explorers. The ever-efficient New South Wales Surveyor General, Thomas Mitchell, even used the abundance of camp-fires across a valley as a default map for the course of the river.[4] Gregory Blaxland in 1813[5] frequently noted the presence of camp-fires, but such smoke was not seen as anything sinister or unusual, as can be seen from John Oxley's description of the Hastings River valley north of Sydney in 1820:

> Numerous smokes arising from natives fires announced a country well inhabited, and gave the whole picture a cheerful aspect, which reflected itself on our minds: and we returned to the tents with lighter hearts and better prospects.[6]

Such references to fire are nothing out of the ordinary and cannot be taken as evidence of either bushfires, or that Aboriginal people were using fire as a widespread way of managing the landscape and its resources. These accounts merely refer to human habitation, the same as might be found on any other part of the globe occupied by people. It is not enough for fire merely to be mentioned by explorers; we must also look for their descriptions of the fires, and their interpretations of what types of fires they observed.

SOMETHING EXTRAORDINARY

The first sign of something out of the ordinary comes from the very earliest written accounts of Australia by a European. In Tasmania in 1642, Abel Tasman noticed

> many trees, which were deeply burnt a little above the ground. The earth here and there had been worked by hand and baked as hard as flint by fires. A short time before we got sight of our boats returning to the ships, we now and then saw clouds of dense smoke rising up from the land, which was nearly west by north of us, and surmised this might be a signal given by our men ... When our men had come on board again, we inquired of them whether they had been there and made a fire, to which they returned a negative answer, adding, however, that at various times and points in the wood they also had seen clouds of smoke ascending. So there can be no doubt there must be men here of extraordinary stature.[7]

As they sailed away from the Tasmanian coast two days later, Tasman reported seeing much smoke arising through the day. This description seems to be more than just the smoke from campfires. The observation of trees burnt around their bases certainly signifies a bushfire, as does the earth baked hard as flint. While the observations of smoke might be nothing more than camp-fires, Tasman's comment about men of extraordinary stature suggests that he was witnessing something bigger than he might have expected.

Members of the First Fleet in 1788 similarly observed a large fire 'made by natives' at Storm Bay in Tasmania on their way to the east coast. On landing, the surgeon John White noticed a burning tree that had apparently been set on fire by lightning. Such lightning- and fire-damaged trees appeared to be common. White also noticed 'that every part of the country, though the most inaccessible and rocky, appeared as if, at certain times of the year, it had been all on fire'.[8]

On the other side of the country, similar observations were made by George Vancouver in King George Sound.

> ... the very extraordinary devastation by fire which the vegetable production had suffered throughout the whole country we had traversed ... we did not see a spot ... which had not visibly felt its effects. Where the country was well-wooded the loftier timbers had their topmost branches burnt, yet none seemed totally destroyed.[9]

Fires certainly added to the difficulties of establishing an English settlement in Australia. The English had brought little in the way of horticultural or agricultural expertise with them and struggled with the basics of growing crops in this very different climate and conditions. Having managed to grow a few meagre crops, the new settlers were hit by drought, desiccating their first attempts at self-sufficiency. Even those who managed a harvest were not spared, for fierce winds set fire to the grass and trees with a ferocity that astonished Governor John Hunter. All hands, even those of the governor, were called to battle a fire that threatened wheat in Parramatta. 'The night was dark, the wind high, and the fire, from its extent, and the noise it made thro lofty blazing woods, was truly terrible,' Hunter recalled, but he was confident that 'as we clear and lay open the country we shall get the better of such accidents'.[10] Hunter ordered firebreaks be ploughed around settlements to protect them from such fires, which he realised were often set by both escaped convicts and Aboriginal people.

UNDER FIRE

Perhaps not surprisingly, given normal Aboriginal-colonial interactions, it did not take long for descriptions of fire in the Australian landscape to shift to hostility. On 19 July 1770, James Cook was careening his ship for repairs at Endeavour River in north Queensland. They were visited by a small party of locals. Cook's crew had killed some turtles and the local men demanded their share. Cook refused, offering them bread instead, and the party left in disgust but not before setting fire to the long grass around the Europeans' camp onshore. Joseph Banks was taken by surprise at how fiercely the grass burnt in the hot climate, and noted that in future he would take care to burn the surrounding areas before setting camp.[11]

The antagonistic use of fire was not confined to English-Aboriginal relations. French scientific expeditions received the same treatment. In 1802, the French naturalist François Péron was astonished to observe

> the great number of fires that we could see. On all sides there rose great columns of flame and smoke; over a stretch of several leagues, all the slopes of the mountains at the head of the North West Bay were on fire ... Thus perish these ancient and venerable forests, which Time's scythe spared over several centuries only to deliver them up more untouched to the destructive instinct of their savage inhabitants.[12]

A few days later, Péron again encountered the use of fire by Tasmanians and this time it seems clear that its function is hostile.

> ... swirling black smoke rose on all sides, every part of the forest was on fire; its wild inhabitants seemed to want, at all costs, to send us away from their shores. They had withdrawn to a high mountain which, itself, looked like an enormous pyramid of flame and smoke, from there, their clamour could be heard, and the gathering of people appeared to be large ... we headed west towards the burning mountain just mentioned. It was a horrible sight: the fire had destroyed all the undergrowth; most of the bushes and shrubs had suffered the same fate; the bigger trees were burnt to a very great height; in some places the fury of the flames had toppled them and their remains had created vast, burning masses.[13]

Since the intruders clearly failed to take the hint, the Tasmanians tried the same tactic the next day, with similar lack of success.

These early European intruders must have seemed positively obtuse to the Aboriginal inhabitants. Despite having been sent the clear and unavoidable message that they were unwelcome, they not only refused to leave but pressed forward, often seeking out contact with the very people who were doing their best to deter them. Having observed a pair of Aboriginal men carefully setting up a decoy fire to cover the tracks of their retreating tribe, Thomas Mitchell in 1846 appears quite unperturbed by this unflattering reception, noting only that the dense smoke added to the sublimity of the sunset, which he sat down to draw.[14] Mitchell was frequently the subject of Aboriginal attempts to burn him out, noting it to be one of the first hostile strategies used.

Aggressive use of fire in warfare was observed even in the wet tropics of Australia's north-east coast. On Kennedy's ill-fated expedition in 1849, William Carron described the following fearsome encounter:

> Twelve or fourteen natives made their appearance at the camp this evening, in the same direction as on the previous day. Each one was armed with a large bundle of spears, and with boomerangs. Their bodies were painted with a yellowish earth, which with their warlike gestures made them look very ferocious. The grass in the position they had taken up was very long and very dry, quite up to the edge of the gully; they set it on fire in three or four places, and the wind blowing from them to us, it burned very rapidly. Thinking we should be frightened at this display they followed the fire with their spears shipped, making a most hideous noise, and with the most savage gestures.[15]

HUNTING BY FIRE

Aboriginal fondness for fire was well evident to early European settlers. Arthur Phillip described their 'perpetual fires'[16] as bordering on a national religion. Ernest Giles, who explored areas of central South Australia and Western Australia in the 1870s, famously declared 'the natives were about, burning, burning, ever burning, one would think they were of the famed salamander race, and lived on fire instead of water'.[17]

While Giles was frequently on the receiving end of hostile fires, he also appreciated the diversity of uses to which fire was put, including signal and hunting fires. According to Giles, Aboriginal people rarely hunted without using grass fires, and that extraordinarily large tracts of land might be burnt in this way. David Carnegie in 1896 also observed the use of fires in the West Australian desert for signalling, burning off old spinifex to promote new growth (and thus lure grazing animals) or fires to flush out rats and lizards.[18]

Aboriginal hunting fires were frequently observed and described in Tasmania. Just such a scene was immortalised by the convict artist Joseph Lycett in *Aborigines using fire to hunt kangaroos* in 1818. The missionary George Robinson describes a similar event on Tasmania's north-west coast in 1834.

> The underwood was very dense and swarmed with mosquitos. The natives commenced burning the bush and in a few minutes the woods were in one general blaze. The lurid flames sending forth volumes of thick smoke had a grand and imposing appearance. The natives fire the wood to drive up the game. The animals run frequently through the flame or jump through it.[19]

Western Australia too, provides clear evidence of hunting fires. Scott Nind, a medical officer stationed at King George Sound, provided detailed observations of Aboriginal burning and hunting practices. His observations conform with those of John Stokes, who described the following scene a few years later in 1841:

> On our way we met a party of natives engaged in burning the bush, which they do in sections every year. The dexterity with which they manage so proverbially a dangerous agent as fire is indeed astonishing. Those to whom this duty is especially entrusted, and who guide or stop the running flame, are armed with large green boughs, with which, if it moves in a wrong direction, they beat it out. Their only object in these periodical conflagrations seems to be the destruction of the various snakes, lizards, and small kangaroos, called wallaby, which with shouts and yells they thus force from their covert, to be despatched by the spears or throwing-sticks of the hunting division. The whole scene is a most animated one, and the eager savage, every muscle in action and every faculty called forth, then appears to the utmost advantage, and is indeed almost another being. I can conceive no finer subject for a picture than a party of these swarthy beings engaged in kindling, moderating, and directing the destructive element, which under their care seems almost to change its nature, acquiring, as it were, complete docility, instead of the ungovernable fury we are accustomed to ascribe to it. Dashing through the thick underwood, amidst volumes of smoke—their dark active limbs and excited features burnished by the fierce glow of the fire—they present a spectacle which it rarely falls to our lot

> to behold, and of which it is impossible to convey any adequate idea by words.[20]

It is clear then, that Aboriginal people deliberately used and to some extent, controlled fire for hunting. What impact did that use of fire have on the landscape?

VEGETATION MANAGEMENT

The fervent George Robinson made it his mission to collect up the remaining 'wild' Tasmanian Aboriginal people and take them off to the relative safety of Flinders Island. Robinson trailed about the west coast of Tasmania between 1829 and 1832 with an entourage of porters and friendly Aboriginal people, making contact with remaining tribes and persuading them to give up their traditional ways and come under his stewardship. Robinson frequently mentioned the occurrence of burnt areas of vegetation in his copious journals. He assumes that this burning has been by Aboriginal people to keep areas open for travel and also notes that recently burnt areas are good for hunting and particularly favoured by the 'boomer and forest kangaroo' we know as eastern grey kangaroos.[21]

Robinson's journals are often cited as providing proof that west coast Aboriginal people maintained the sedge and heathlands of their coast, while protecting small groves of trees from flames.[22] In this wet environment, where lightning fires occur less frequently, it is certainly likely that Aboriginal burning played an important role in maintaining open spaces. Robinson rarely appears to have witnessed this activity himself and hence his journals provide little direct evidence of Aboriginal burning other than hearsay.

The view from the south

Imagine that you are standing on a beach on the south coast of Tasmania, looking across nearly 3000 kilometres of empty ocean that separates us from Antarctica. From here you can see Maatsuyker Island, just beyond the De Witt Islands. Further out, you can see the Mewstone, the characteristic rock sought out by early European sailors following in the wake of Abel Tasman. The sight of this rock, after weeks, months, of empty seas, symbolised the prospect of safe harbour in an isolated wilderness on the opposite side of the world from their own.

Lashed by the fury of the Southern Ocean, the rocks here seem bare and grey, an inhospitable place for life. Rain falls on 250 days of the year. But in defiance of salt and spray, a thick carpet of vegetation creeps down the rocks. The soil is rich, manured by countless generations of muttonbirds, whose mournful cackles echo up from the honeycomb of burrows all over the islands.

These southern islands seem a strange place to consider fire in Australia. Yet fire leaves its mark even here in this damp, cold extremity. When Matthew Flinders sailed past the De Witt Islands on 13 December 1798, he was surprised to notice that they had been recently burnt.[23]

To the north lies the main land mass of Tasmania. From the sea, the coastline looks much the same as it did 300 years ago when Europeans first sailed across the horizon. In fact, it probably looks much the same as it did thousands of years ago when Aboriginal people first crossed the land bridge connecting Tasmania to mainland Australia. Along the south and up the west coast, the Tasmanian wilderness remains as inhospitable to humans now as it was then, a tangled mass of vines, huge old trees and fallen branches, crumbling with decay and bursting with new life. 'The beauties of unspoilt nature', wrote one early explorer of this wilderness, 'the one of always being old, and always new'.[24] The appearance of antiquity is not deceiving. These are ancient forests indeed.

In one corner of the south-west wilderness lives a plant known as King's holly or King's lomatia. This low-growing shrub, with characteristic dark serrated leaves and red berries, is the only known member of its species. It lives alone in the world, slowly spreading by suckers, ever expanding its extent. This single vast organism now covers over 1 square kilometre and appears to have been growing here in splendid isolation for 43 000 years.

Further north, on the shores of a glacier lake near Mount Reid, grows a small forest of Huon pines. All are male and all are genetically identical. They are clones, all grown from the one plant that seems to have germinated here at least 10 000 years ago, leaving its pollen embedded in the sediments of the lake. Surrounding this ancient colony are other conifers: billy pines, pencil pines, chestnut pines, creeping pines and celery-top, some of which are over 1000 years old themselves.[25]

These ancient plants grow in the forest remnants of the Antarctic beech–southern conifer assemblage that once spread across southern Gondwana 100 million years ago. Today its descendants survive only the damp, cold places of the sub-Antarctic islands, New Zealand and Tasmania. They survive only where there is no fire. A single fire would destroy these ancient conifer forests, perhaps forever.

Tasmania is not the only home to such fire-sensitive antiquities. Ancient remnants of the long-distant conifer forests also find refuge in the deeper, damper gullies of the mountains of New South Wales—like the 80 or so Wollemi pines that cluster in a single deep gorge, their dark fishbone leaves radiating out from chocolate-brown trunks of bubbled bark. One of the rarest plants in the world, these prehistoric pines stand out from the tree ferns and more modern flowering coachwoods and sassafras that grow around them in the damp conditions. Whatever the reason for their survival here and nowhere else, the absence of fire has certainly been an essential precondition.[26]

By the time Ludwig Leichhardt explored inland Queensland in the 1840s, European expansion in the eastern hinterland was well underway. The fire practices he described appear common among both Aboriginal people and Europeans alike, so it is unclear who was influencing whom.

> This grass ripens in October and November, when the ground under the trees looks like an even, sweeping field of oats. In November and December the weather gets dry and the bushfires break out. They're often a mile wide, and they clear the ground of grass and dead wood as they sweep through the bush. The ashes left behind are like manure to the sweet, tender grass that shoots up as soon as it rains and is eagerly consumed by kangaroos, flocks of sheep and herds of cattle.

> During the hot months fire can range over hundreds of miles. It starts where the blacks have been camping for the night, as all they do when moving on is to pull a brightly burning stick out and keep it smouldering against a piece of bark, so that they can light a fire at the next camp; or white men have set the dry grass on fire passing on; or settlers have been systematically burning off the old grass to obtain young pasture for their sheep and cattle.[27]

Thomas Mitchell, on his regular and detailed explorations of the inland areas of New South Wales, south to Victoria and north into Queensland, reported extensive grass fires and forest fires, often noting the dryness of the countryside and how easily it might ignite.

> We approached this position with our carts, in the midst of smoke and flame; the natives having availed themselves of a hot wind to burn as much as they could of the old grass, and a prickly weed which, being removed, would admit the growth of a green crop, on which the kangaroos come to feed, and are then more easily got at.[28]

CONTROLLED BURNS

Even with modern firefighting equipment, it is difficult to control fire. No doubt Aboriginal Australians sought to control fire to prevent it burning sensitive areas, but it is unlikely that they would always have been successful. Many observations suggest a more casual approach to fire than a strict model of 'fire-stick farming' or controlled burning might suggest.

Edward Eyre commented on the 'casual fires so frequently left by them on their line of march'.[29]

There seems little doubt that Aboriginal Australians frequently and deliberately started fires, either in warfare or for hunting, but it is less obvious how much control they had over these fires once they started. In the southern regions, fires are easiest to control in autumn and in spring, when the vegetation is dry enough to burn but not so dry that it burns out of control. These times of year are often the best times to promote fresh seasonal growth. And yet a review of historical records of Aboriginal fires found that nearly three-quarters were lit in summer, and most occurred in the hottest four months of the year under typically hot and windy conditions.[30] This may reflect a large number of hostile fires rather than more controlled seasonal burns. Either way, it reveals that fire is at best a lightly reined beast and a servant that all too frequently explodes out of control.

In the northern areas, the peak bushfire period is during the dry months from winter to spring. Henry Keppel, on his visit to Port Essington in the Northern Territory in 1853, wrote:

> About the middle of the dry season, the natives set fire to the grass which is abundant everywhere, and at that time quite dry; they do it annually, in order to clear the country for their wanderings, to destroy vermin, and to promote the growth of roots which they require as articles of food. The conflagration spreads with fearful rapidity and violence... until the whole country as far as the eye can reach, is in grand and brilliant illumination.[31]

It must have been observations such as Keppel's that persuaded so many of the early explorers, like Mitchell and Giles, that

Aboriginal people burnt the landscape deliberately and with great effect. It must have been this kind of behaviour that led Mitchell to conclude that there was a codependent relationship between fire, grass, kangaroos and humans in Australia. Mitchell was in no doubt that Aboriginal people deliberately used fire to remove unwanted vegetation, promote fresh grass growth and encourage game. Like Giles, he witnessed the use of fire to hunt small mammals. Like Leichhardt, he saw the effects of fire on the vegetation over too many years and in too many places to feel that the changes were coincidental.

It was clear to these early explorers, these early contact 'anthropologists', that Aboriginal people used fire deliberately and effectively to manage resources. Similarly, Edward Curr in Victoria, like other early settlers, described his Aboriginal neighbours as 'living principally on wild roots and animals, he tilled his land and cultivated his pastures with fire'.[32] Curr, incidentally, didn't think this was a particularly productive method of agriculture and attributed the often barren nature of the Australian landscape to Aboriginal burning.

Fire was undoubtedly used deliberately by the original inhabitants of Australia, for hunting and travel and probably to promote new growth. But did this use of fire change the Australian landscape? For Mitchell, the gradual thickening of the forests and undergrowth around Sydney could only be due to the cessation of traditional burning, to the change from Aboriginal land management to European land management.[33]

But just how real was this change from open woodlands to dense thickets? Or was there something else going on at the same time that could account for that change?

4 · A WALK IN THE PARK

During the hot dry weather of the Christmas time, very extensive bush fires spread about the country, and were sometimes extremely mischievous in their destruction of fences, which are very liable to be thus burned, unless care be taken, previously to the dry season, to clear away all fallen wood and rubbish, and to burn the high grass and ferns for a breadth of three or four yards on either side. The fences of sheep-runs, which extend in lines of many miles in length, over the uncleared hills and forests, are those which most frequently suffer; but growing crops, stacks, farm-buildings, and dwellings are likewise sometimes swept away by the rapidly advancing fire.

By day, the effect of these great conflagrations was far from pleasant, causing an increase of heat in the air, and a thick haze over the landscape generally; whilst from the various points where fires were raging, huge columns and clouds of dense smoke were seen rising as if from volcanoes: but at night, the scene was often very grand; sometimes the fire might be watched, on any rising ground, spreading onwards and upwards, swifter and brighter as it continually gained strength, till the whole mountain side was blazing together; and after the first fierce general flames had passed away, and the great trunks of trees alone remained burning, the effect resembled that of the scattered lights seen on approaching a distant city at night. The rocky Schoutens glittered with partial lines and trains of fire, that marked their rugged and lofty outline like burnished gold amidst the darkness. Each night showed some new change in the great illumination, until a heavy fall of rain extinguished it altogether, much to the satisfaction of all who feared its nearer approach.

A recent scientific writer (the Count Strzelecki) in treating of this colony, condemns the practice of burning, as seriously injurious to the pasturage, and seems to suppose that the custom originated with the colonists; whereas the aborigines practised it constantly, knowing the advantages of destroying the dense growth of shrubs and coarse plants which cover the country in many parts, and spring up again after the fire with young and fresh shoots, which many of the wild animals then gladly feed on. The grass also grows again immediately after the fires, and is greatly preferred by all animals to the old growth; whilst, from the destruction of tall ferns and scrub, it is rendered more accessible to them. Sheep-owners know how serviceable occasional bush fires are, and generally arrange to burn portions of their sheep-runs at different times, so as to have a new growth about every three years. Where this is neglected for a length of time, the rank luxuriance of the great brake fern and other uneatable plants, and the accumulated mass of dead wood, bark, and leaves, form such a body of fuel, that when a fire does reach it, the conflagration is thrice as mischievous in the destruction offences as it otherwise would have been.

Louisa Meredith 1856[1]

Climate change may well drive the major shifts in Australia's vegetation, but Aboriginal burning does appear to have exerted some local effects on particular patterns of vegetation in particular areas. How big those changes were, and whether they are good or bad, remains contentious.

What is less contentious is that Australia's fire regime changed again with the advent of European settlement. Some argue that Europeans have decreased the frequency of fires, increasing their severity once they do occur. Others argue that early Europeans were pyromaniacs who burnt the landscape uncontrollably. Australia's vegetation certainly changed dramatically through widespread clearing with European settlement, but have our few remaining areas of native bush also changed? If we want to restore or retain them in their natural state, what role should fire play?

ROLLING PLAINS AS FAR AS THE EYE CAN SEE

Open vistas of rolling plains, dotted with picturesque trees and the occasional watercourse, are almost universally popular landscapes to the human eye. Perhaps they are reminiscent of our savannah heritage. Such landscapes promise easy passage, good hunting and security from ambush predators for a vulnerable, two-legged primate travelling in small family groups and dependent on good eyesight and quick wits for safety.

So it is perhaps not surprising that we should seek out such landscapes, particularly when these landscapes are also those most suited for our favoured system of agriculture. Humans are grass-eaters. Grass is our dominant food supply. Grasses, and the grains they produce, account for half of all the kilojoules consumed by humans.[2] Grasslands feed us, both directly and indirectly, through the stock we graze on them. Naturally we seek out grasslands and comment on them when we find them.

The explorer James Cook described just such a landscape of open woodland in Botany Bay on first arrival.

> ... the woods are free from underwood of every kind and the trees are at such a distance from one another that the whole country or at least a great part of it, might be cultivated without being oblig'd to cut down a single tree; we found the soil everywhere except in the marshes to be a light white sand and produceth a quantity of good grass which grows in little tufts ...[3]

John Batman described a similar landscape in Port Phillip.

> I went on shore to look at the land, which had appeared beautiful, with scarcely any timber on. On my landing I found the hills of most superior description beyond my most sanguine expectations. The land excellent and very rich—and light black soil covered with kangaroo grass two feet high and as thick as it could stand. Good hay could be made, and in any quantity. The trees were not more than six to the acre, and those small sheoak and wattle. I never saw anything equal to the land in my life. I walked over a considerable extent, and all of the same description. This land forms an isthmus which is about 20 miles [32 km] long by 10 across it—upwards of one hundred thousand acres or more. I could see five or six miles in every direction. Most of the high hills were covered with grass to the summit, and not a tree, although the land was good as land could be. The whole appeared like land laid out in farms for some one hundred years back and every tree transplanted. I was never so astonished in my life ...[4]

As it happens, the farmer Batman's account proved more accurate than that of the mariner Cook. The First Fleet hastily abandoned Botany Bay for the more promising environs of Port Jackson (complaining that no-one could find Cook's 'meadows'),[5] while Batman's selection at Port Phillip proved the basis for Melbourne, and a thriving agricultural industry.

The abundance of these descriptions of open grasslands in early accounts of the Australian landscape is often taken as proof that fire was an important feature of the pre-European ecosystem. Fire is often thought to be the major way in which grasslands are maintained in the face of encroaching woodland and forests. While grasslands are often favoured by fire, other factors also promote low, unwooded vegetation types, most notably lack of water, low nutrient soils and grazing. In Africa, for example, it is thought that elephants play an important role in maintaining the savannahs by pushing over trees. Lack of undergrowth in an open forest and a grassy understorey is also a common feature of many mature, undisturbed eucalypt forests. Fire is not the only creator of grasslands.

IMPENETRABLE SCRUB

We should also be wary of drawing too much from the accounts of explorers and settlers in their descriptions of vegetation. Those explicitly charged with looking for pasture will naturally be inclined to write about promising locations, rather than unpromising ones. It is not surprising that the Surveyor General of New South Wales, Thomas Mitchell, should have described open grasslands in parts of Victoria. What would be surprising is if he described nothing else. Vegetation changes dramatically even across small differences in soil and topography. Thickly vegetated gullies naturally abut sparsely wooded plains interspersed with open hilltops.

As an example, the French explorer Dumont d'Urville, more interested in botany than colonies, describes the eastern shore of Western Port, in the vicinity of Bass River, as being 'lovely grassland ... shaded with fine trees ... rather like our royal forests around Paris'.[6] But spend a little more time with Dumont d'Urville and we soon discover that on the opposite side of the anchorage, on Phillip Island, the vegetation is dense and utterly impenetrable. Similarly, even a cursory glance at Mitchell's journals reveals an abundance of impassable swamps, formidable 'malga' scrub and close-growing brigalow, along with fine pastures and open woodlands.[7]

There is little doubt that parts of Australia at the time of European settlement were furnished with open woods and grasslands. Fire, both natural and deliberately lit, probably played a role in maintaining them. There is also little doubt that large parts of the country were carpeted in a wide variety of other vegetations, many of which are highly sensitive to fire.

More interesting is the observation of many early settlers and explorers who felt that those much-prized open woodlands and grasslands were disappearing, even in the short time they had been in Australia.

Mitchell argued that the once-open forests of the Cunningham Plains in New South Wales had been rendered impassable by thick impenetrable undergrowth that had grown since the cessation of annual burning by Aboriginal people.[8]

Botanists think Mitchell was probably right in this case.[9] Many understorey species grow back thick and fast after fires, logging and clearing, unless clearing activity is maintained. Such regrowth would be characteristic of both an absence of fires in this ecosystem or a response to one-off clearing or logging, which is also likely to have occurred at this time. Mitchell's description of being able to gallop a horse through the forest of the Cumberland Plains, and the regrowth of the area, probably does accurately reflect changes in the vegetation being caused by lack of burning. But this does not mean that all parts of Australia were open wooded plains, or that all plains like this were maintained by burning. The open, shrubby heathlands of Sydney's eastern suburbs were more likely created by the low-nutrient sandstones upon which they grow, rather than fire.

DISTURBING EVIDENCE

It has taken decades to gain an understanding of how Australian vegetation responds to disturbances, including fire. For many years, colonists stuck with European and North American models of forest development, which they tried to apply to Australia's dramatically different vegetation. In Europe and North America mature forests are often seen as stable, 'climax' communities that remain in their mature form for centuries, undergoing a slow gradual process of renewal that barely changes their external appearance overall. This model fails miserably to explain Australian forests, which are better described in terms of continual successional cycles of destruction and renewal. Small wonder early settlers puzzled over the strange behaviour of these alien forests.

Alfred Howitt describes just such a strange successional process in the Gippsland area of Victoria in 1890, where gigantic, dead black ash trees were surrounded by an impenetrable thicket of saplings, only some 12 years old.[10] Black ash, like the more famous mountain ash, typically regenerates prolifically after fire, grazing or logging. In this case, regeneration appears to have been triggered by the death of the mature trees that previously shaded the ground, restricting germination. Far from being evidence of a failed fire regime, this observation seems to describe a fairly typical ash forest response to fire or some other disturbance. Left to its own devices, it would revert back to an open stand of mature trees over the course of several decades.

Howitt was no mere ignorant passer-by. He spent years studying and documenting both the natural history and Aboriginal people of the Gippsland area. In a paper on Victorian eucalypts presented to the Royal Society of Victoria, Howitt noted:

> I might go on giving many more instances of this growth of the Eucalyptus forests with the last quarter of a century, but those I have given will serve to show how widespread this re-foresting of the country has been since the time when the white man appeared in Gippsland, and dispossessed the aboriginal occupiers, to whom we owe more than is generally surmised for having unintentionally prepared it, by their annual burnings, for our occupation.[11]

This concept, that Europeans had increased the forested area of Australia, certainly went against the prevailing view of members of the Royal Society that the country was being denuded.[12] But we must recall that Howitt was referring to one particular ecosystem and that what applies for one forest type will not necessarily apply in another. In the dense, wet forests of Gippsland, fire would need to be deliberately applied to maintain open areas, much as it was in the dense, wet forests of the west coast of Tasmania. Here, James

Walker, in 1889, took a very similar view to Howitt:

> It appears that the blacks were accustomed to take considerable pains, by means of periodical burnings, to keep down the scrub and promote the growth of grass on their favourite hunting grounds. Many open plains, especially in the north, which were formally known as favourite resorts of the blacks, subsequently became overgrown with forest through the discontinuation of these annual burns.[13]

Much clearer evidence of the impact of Aboriginal burning practices comes from Western Australia, where the impact of European colonisation was later and somewhat less destructive of local culture. In 1818, Phillip Parker King noted that:

> a lapse of sixteen years will in this country create a complete revolution in vegetation; which is here so luxuriant and rapid that whole woods may have been burnt down by the natives, grow again within that space of time.[14]

Frequent fires certainly appear to have been an important part of the ecology of south-west forests and heathlands for a long time.

The burning bush

The Tasmanian blue gum is, in many ways, the archetypal gum tree. It was the first eucalypt to be named and described, by the French botanist Labillardiere in 1791. It gives us Australia's only native medicinal export—eucalyptus oil. And despite its restricted native distribution (to Tasmania and isolated coastal pockets in Victoria), the blue gum has been a highly successful export—to Western Australia, California and China. It is the most widely planted plantation tree in the world.

Blue gums are rarely killed by fire. Thick bark insulates the base of the trunk before stripping off in long stringy shreds. In the canopy, the oils in the leaves rapidly evaporate in the heat, adding their quick-burning gases to the volatile mix. Yet despite the sudden fierceness of a blue gum in flame, the fire rarely lasts long and rarely generates enough heat to cause any lasting damage to the tree. Within days of a fire, new shoots emerge along the trunk. Within weeks, the once-blackened tree stump resembles a green woolly monster, encased in the light, bright, juvenile leaves rapidly maturing to their characteristic blue-grey colour.

In some cases, an abundance of leaf litter and undergrowth might produce a fire intense enough to kill the dormant buds on the trunk of the blue gum. But, like the vast majority of eucalypts, blue gums swell just beneath the ground to form a bulbous lignotuber. This underground larder provides food for the plant when its upper limbs are damaged, to allow regrowth. Should the top of the tree be destroyed, the dormant buds on the lignotuber surface reshoot, allowing the undamaged root system to spring back to life in a coppice of small trunks. Like most eucalypts, blue gums are fire-tolerant trees. They don't need fire, but they can survive it. Across Bass Strait, in Victoria, another giant of the forests depends on fire for its very survival.

In a mature mountain ash forest, there are no

babies. Young mountain ash cannot grow in shade. In the shadow of their elders, there is no prospect of survival. The uniform pillars of mountain ash trunks are all the same age—a generational cohort of peers all growing, maturing and ageing in unison. On clear ground, they sprout rapidly and vigorously, spreading out a thick, impenetrable mat of roots, excluding any other plants from sharing their space. Crammed together—2.5 million per hectare—the young saplings race skyward, the tallest slowly shading and killing their smaller and slower rivals until eventually, after a century, only 20 giants per hectare remain. Beneath them, shade-dwelling plants like wattles, tree ferns and other rainforest plants find a home.

As centurions, the mountain ash are at their prime of life and they begin the slow decline into old age. They shed seeds into the deep leaf litter, but in the cool, dark shade, the seeds lie dormant, unable to germinate. Branches crack and crash to the forest floor. Hollows form where limbs once grew, forming homes for possums and birds. The canopy recedes, thinning to reveal skeletal branches. Long ribbons of bark stream down their trunks, collecting in great drifts at their bases. In another hundred years, the old giants will begin to fall. Patches are covered by the rainforest and scrubby undergrowth. The mountain ash forests, and the animals dependent on them, disappear.

The mountain ash plays a game of chance. Fire is not by nature a frequent visitor to these cool, moist forests. Once every few centuries, the conditions are right—a hot summer, a long drought, dry summer thunderstorms. A crack of lightning in the Alps lights the tinder-box forest and the fire erupts, feeding on the mountain of accumulated fuel around the trunks of the ancient trees, growing into an all-consuming ferocious beast.

These are hot, fierce, severe fires. Some estimates rate their output at 60 to 100 kilowatts of energy.[15] Fire ecologist Kevin Tolhurst estimated one fire to have produced enough energy to fuel 1500 atomic bombs. The mountain ash offers no resistance to the inferno. Unlike other eucalypts, they don't store reserves in underground lignotubers or protect delicate buds beneath thick bark. In a fierce fire, they simply die—releasing in their death throes a shower of seeds from their hard woody seedcaps. And on the scorched nitrogen-rich earth below, a new generation of mountain ash rises to take the place of its predecessors. It's a risky strategy. Without fire, there is no mountain ash. Too many fires will kill the new saplings before they can seed, and the mountain ash will disappear also. Like many fire-dependent species, mountain ash is highly sensitive to changes in the rhythm and pattern of fire it has adapted to over millions of years.

SMOKE IN THE GRASS

Sometimes the answers to these questions are engraved into the hearts of the trees themselves. Long-lived grass trees are renowned for their ability to regenerate after fire, often flowering prolifically in the blackened landscape. Coloured banding at the bases of their leaves reveals the history of fire over their lifetimes, often as long as 250 years. David Ward and his colleagues at Curtin University of Technology found that the frequency of burnt layers in grass trees from the jarrah forests of south-west Western Australia has decreased steadily since 1750 from an average of one fire every three years to fewer than one every 10 or 15 years. Ward attributed this steady decrease to the reduction in Aboriginal burning practices, particularly from the mid-1850s, when a sudden decline is most apparent.[16]

FIRE AND SNOW

While fire seems to have decreased since European settlement in some areas, there is evidence it has increased in others. Snow-gum woodlands are scattered across a mosaic of grassland and alpine heath, creating the picturesque landscape immortalised by Elyne Mitchell and Banjo Paterson. Today the open woodlands burn frequently here, with cool, quick burns killing off shrubs that might grow into a dense understorey. Arboreal animals like koalas and possums climb up into the high canopy to escape the flames. Wombats and bandicoots flee to cool dens below the earth. Fleet-footed animals use their speed to escape. Rocky outcrops and the wet sphagnum moss bogs of the alpine regions provide shelter for other species.

After a cool fire, the grasses and other ground-cover 'fireweeds' regenerate rapidly. The mature snow gums resprout from the epicormic buds on their trunks and branches or from their underground tubers.

Without frequent fires, the landscape changes dramatically. Saplings and shrubs thrive in the understorey, thickening and filling the available space, creating dense clumps of vegetation around the mature trees. With the increase in fuel, fires now will burn hot and fierce around the trunks, killing the snow gums outright. These fiercer fires can cause lasting damage in drought years when the bogs dry up, allowing the layers of peat below the surface to catch fire, destroying ecosystems that have taken centuries to develop.[17]

And yet despite the appearance that cool frequent burns are ecologically appropriate, the tree rings suggest that fires have increased since European settlement. Fires seem to have been infrequent (one to two large fires per century) in the alpine snow-gum woodlands before European settlement but suddenly increased with the introduction of European settlers and their livestock to the alpine and subalpine areas of south-eastern Australia. A study of tree rings in the Snowy River valley found no evidence of any fires at all between 1750 and 1845. From 1845 onwards, fires occurred frequently, every six years or so. A variety of studies of different tree-ring patterns suggest that there was little Aboriginal burning in the Alps, but that European grazers brought fire with them, increasing the incidence of fire sevenfold.[18]

Did Aboriginal peoples burn the land or not? Have Europeans increased, or decreased, the frequency of fires in Australia? If we wish to maintain our forests, grasslands and woodlands in any kind of healthy state, should we be burning or suppressing fire?

The trouble lies not with the answers but with our questions. We cannot easily answer questions about the impact of burning on the Australian landscape because there is no single 'Australian landscape' any more than there is a uniform representative form of either 'Aboriginal burning' or European burning. The impact of burning can only be assessed in relation to each ecosystem, each region, each culture and each community. Fire in the Australian landscape is inherently specific to its location. The grass fire of the tropical savannah is a very distant relative of the forest fire of the south. It makes no sense to compare them.

But, within the context of particular communities, both human and plant, there is evidence that many Aboriginal people opened forests, promoted grasslands and cleared woodlands. Fire was undoubtedly a widely used tool and it may well have had significant local impacts on the landscape. But the impact is just that: local. There is little evidence of wide-scale changes to the landscape that were unequivocally caused by Aboriginal Australians. The changes that are evident, such as the long-term expansion of the grasslands and the more recent re-expansion of the forests, appear to be due to climatic changes, not human changes.

The difficulties of interpreting the patterns of change in Australian vegetation are well illustrated by a small reserve in Ocean Grove on the outskirts of Melbourne, thought to be an undisturbed relict of a type of open woodlands described in such glowing terms by John Batman on his arrival in Port Phillip. Visit the Ocean Grove Nature Reserve today and you will find small gum trees poking their sparse canopies above a thick brush of black sheoaks. Although the dominant vegetation today, these sheoaks were completely absent from the area a century ago. Botanist Ian Lunt estimates that, while in the early 1800s there were fewer than 20 trees per hectare, today there are more than 3000 trees per hectare. It is likely that the open woodlands described by Batman were maintained at least in part by fires. It seems that, with the cessation of fires in the area, there was a massive expansion of golden wattles that were heavily exploited for the wattle-bark industry. With the cessation of logging and grazing (and fire) in the reserve, sheoaks have since come to dominate, begging the question of what, if anything, constitutes a 'natural' landscape for this area.[19]

If we wish to see what wide-scale changes to the environment look like, we must enter the period of European settlement. For there is no doubt at all that Europeans radically changed the Australian landscape, perhaps irrevocably, in their efforts to tame it, subdue it and dominate it. But there was one aspect of Australia that would not be subdued.

Fire.

II · THE HUMAN IMPACT

5 · A GROWING RISK

RISK AND INCIDENCE

If you look at a fire map of Australia, you would think that fires only occur in the south. A thick band of 'high risk' extends along the south coast, encompassing much of Victoria, Tasmania, part of South Australia and the south-west corner of Western Australia. A more general risk area takes this band further north across the southern quarter of the continent and up the east coast as far as Brisbane. Beyond this, the fire risk declines to moderate, with an island of inland western areas—desert country—being rated as low risk.

But anyone who has been to the northern regions of the country will know that this map is not a map of fire incidence. Fly over the Northern Territory and you will inevitably see fires burning across the open savannahs. In the daytime, clouds of smoke billow in all directions across the horizon. At night, sharp, red lines of fire fronts delineate the landscape for kilometres, appearing, in the distance, like a freeway to the middle of nowhere.

The largest and most frequent fires in Australia occur not in the heat of summer but in late winter and spring: they happen in the north. We may be more familiar with the fierce fires of the densely populated south-east and south-west but 98 per cent of large bushfires occur outside this area. Annual fires in the Northern Territory dwarf those that burn in the southern states—but few people die, so few people care.

Fire risk map

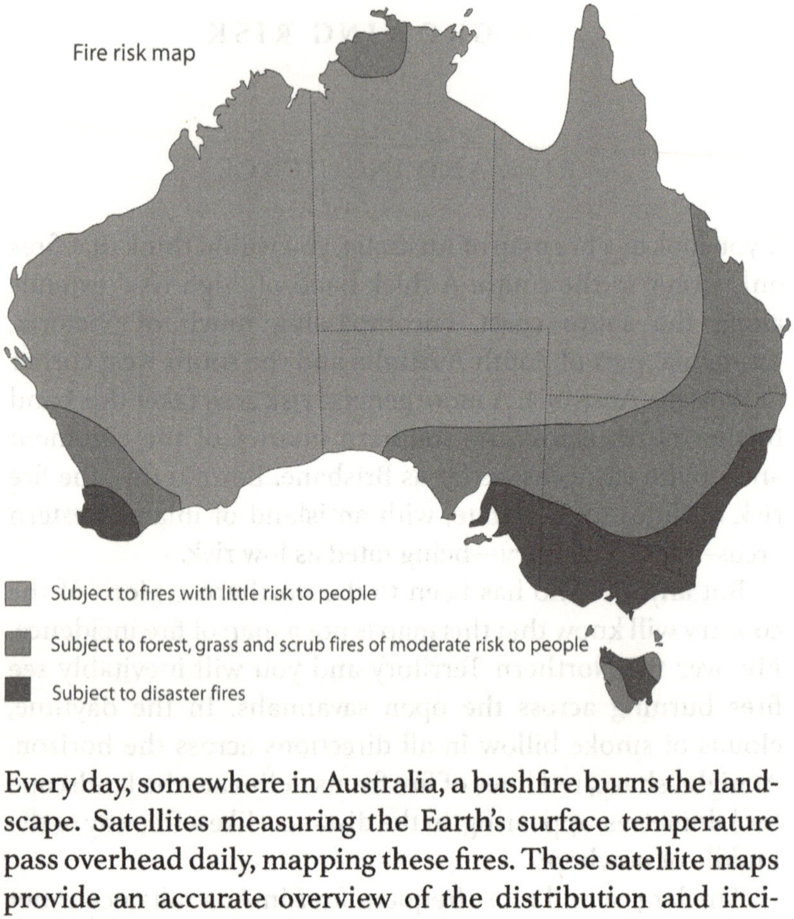

Subject to fires with little risk to people

Subject to forest, grass and scrub fires of moderate risk to people

Subject to disaster fires

Every day, somewhere in Australia, a bushfire burns the landscape. Satellites measuring the Earth's surface temperature pass overhead daily, mapping these fires. These satellite maps provide an accurate overview of the distribution and incidence of fires across the continent and they reveal just how widespread and common burning is.[1]

Winter is the quietest time nationally for fires but, even in this cooler season, fires commonly burn across the northern fringes of the continent, southwards along the east coast into northern New South Wales, in the east of Tasmania and the south-west corner of Western Australia.

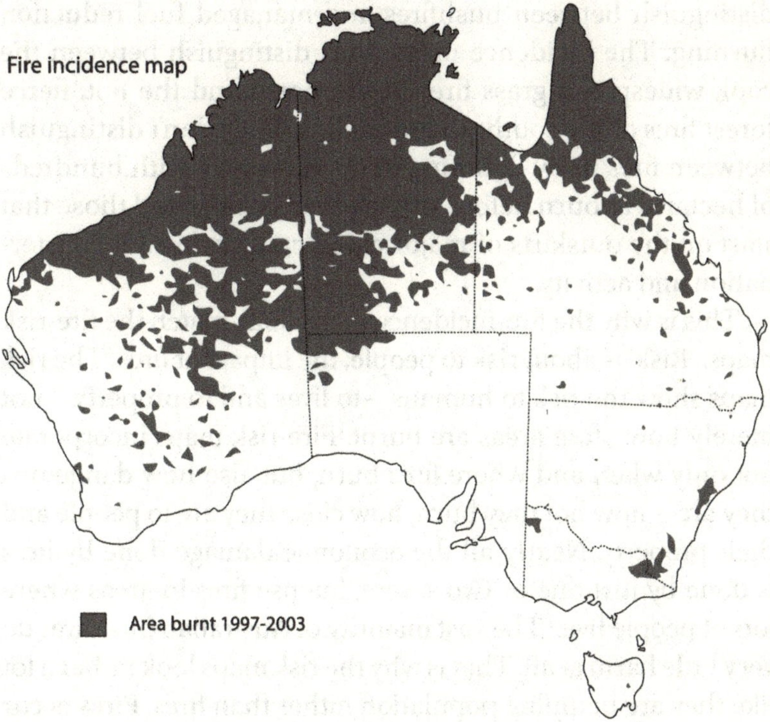

It is only in autumn that this north-south pattern reverses. In the autumn months of March to May, a thick band of fires occurs across the entire south-west corner, most of Tasmania and in a continuous ring from the peninsulas of South Australia, across Victoria and northwards along the coast of New South Wales into south-east Queensland. The line of fires across the north remains, but retreats north to Cape York, into Arnhem Land and the Kimberleys. This southern autumn peak probably reflects the frenetic activity of fuel reduction burning that is crammed into the few cooler months before the winter brings too much rain.

The satellites don't care whether these fires are caused by humans or nature, under control or out of control. They don't distinguish between bushfires and managed fuel reduction burning. The incidence maps don't distinguish between the cool, widespread grass fires of the north and the hot, fierce forest fires of the southern mountains. They don't distinguish between fires lit by lightning in open country with hundreds of hectares to burn before anyone even notices and those that start on the outskirts of major cities, causing instant consternation and activity.

This is why the fire-incidence maps don't match the fire-risk maps. 'Risk' is about risk to people, the impact of fires. The risk maps show the risk to humans—to lives and to property—not merely how often areas are burnt. Fire-risk maps incorporate not only when and where fires burn, but also how dangerous they are—how hot they burn, how close they are to people and their property. Nearly all the economic damage done by fires is done by just one or two severe, intense fires in areas where lots of people live.[2] The vast majority of Australian bushfires do very little harm at all. That is why the risk maps look rather a lot like they are mapping population rather than fires. Fires occur across the continent but people are clustered on the southern and eastern fringes, in the very forests that place them most at risk of the most severe fires of all.

THE PROBLEM WITH PEOPLE

Before European settlement, the population of Australia was probably low, no more than three-quarters of a million people. Even in the most densely populated areas, people were dispersed across the landscape, in proportion to the low-intensity, sustainable system of subsistence that supported them. They moved with the availability and seasonal cycles of food, leaving behind them few permanent structures to be concerned about. For these peripatetic people, closely connected to the subtle changes of their environment, bushfires would simply have been part of everyday life, as much a part of the landscape as snakes, floods or other natural hazards. So while bushfires before European settlement were common, widespread and sometimes severe, human casualties were probably quite low.

With the arrival of Europeans, the pattern of human habitation changed dramatically. Crowded settlements clustered around the major ports like Port Jackson and Hobart Town. Lines of expansion radiated outward from these arrival points, with waves of population density slowly fading as they rippled out from the cramped ports. Closest to the towns were the smallholdings, then the great expanses of the squatters, before the European presence petered out to a mere handful of hardy nomadic individuals—explorers, hunters and prospectors—each on their own voyage of discovery. Newly established ports like Newcastle, Launceston, Port Macquarie, Swan River, Moreton Bay, Albany, Portland, Port Phillip and Adelaide provided further expansion points into the continent. Very soon the European population greatly exceeded that of the rapidly declining Aboriginal population.

Unlike the dispersed Aboriginal people, almost half a million colonists clustered along the richly vegetated areas on the southeastern and south-western corners of the continent.

DREAMS INTO DUST

In Victoria, the open grasslands of the Western District promoted rapid expansion of pastoral leases. From 1835, when

Batman first sent out his encouraging descriptions of pasture from the banks of Port Phillip, a steady stream of sheep had sailed across Bass Strait to the new lands. By the end of 1839 there were over a million sheep, nearly 400 000 cattle and 7000 horses in Victoria. By 1851 the human population of Victoria was nearly 80 000, 60 000 of whom were scattered across the countryside, on vast acreages and struggling farmlets.

All of the best land had been occupied by graziers and farmers. New settlers began to push their way up into more densely vegetated, less accessible areas. The river banks were occupied first, before intrepid souls battled their way through the dense forests to the highland and coastal plains in the far east of the state. It was not a welcoming place. Aboriginal warriors did their best to repel the invaders. Bushrangers preyed on the isolated, far-flung settlements. The very forests themselves seemed to fight back against the farmers who struggled to clear them. As fast as the newcomers ringbarked and chopped down the giants of the forests, a rash of new growth emerged from the soil, smothering crops with their vigour.

Fire proved the easiest way to clear land, an easy way to provide fresh grass for grazing and an easy way to clear a path through the impenetrable forests. Fires lit in the height of summer to clear a small allotment on the edge of a forest burnt merrily on into the wilderness and expanded into great conflagrations, unheeded by the humans who had started them.

By the end of 1850, even the weather seemed to conspire against the settlers in their ramshackle huts. Their crops withered in the heat, waterholes baked hard and dry, and the blistering winds whipped the soil up into great stinging clouds of dust and grit. The losses to livestock were already high. With so little value, and in such poor condition, it was barely worth even getting them to market and many were left to die on the plains. Fires were frequent, burning here and there across the country.

A young Tasmanian schoolgirl, Elizabeth Orr, who had recently moved with her family to the district of Maldon, provided a firsthand account of the situation.

> At the end of January the heat became most oppressive. The sun shone out of a cloudless sky; the grass dry and brown broke crisp and short underfoot. The atmosphere was heavy with smoke and horsemen were arriving from different stations with sad news of the ravages made by bushfires. The thickly timbered lands were burning and the fire creeping down upon the open country, rendering sheep walks and cattle pastures black and desolate.[3]

The station hands had been fighting fires on the outstations for days, trying to protect the grass for the stock. The shepherds' families had been brought into the relative safety of the home station. They all prayed for rain. The fires were an ever-present threat, kept in check by the weather. Rain would bring relief; wind would bring disaster.

The wind arrived before the rain. On the night of 5 February 1851, a hot, fierce northerly set in, reaching gale force the next day. By the morning, temperatures soared to 47°C. The fires could be seen from the cottage where Elizabeth Orr waited. Elizabeth's recollection provides a rare insight into the preparation such rural families routinely took in the event of bushfire.

> There was no time to spare, at the back of the cottage there was a square yard, at the upper end a row of stables and room for harness and conveyances—the day cart that I drove in from Melbourne for one.

> In the stables were some valuable race horses. Other buildings formed the sides of the square—overseer's quarters, men's hut and black smith's forge. At the back of the stables were stacks of hay and straw. The first thing done was to blindfold and lead the horses into the garden in front of the cottage. Mr Simpson had taken the precaution to have the grass burnt in front of the cottage garden when bushfires were at the out stations. The furniture and all valuables with saddles, horse clothing and harnesses were placed in the center of the square yard and covered with wet blankets. There was a good space between the pile of goods and the outbuildings.[4]

When the stables caught fire, the station owner sent his family and young guest to the safety of a hut by the creek surrounded by burnt ground. Even this was no secure bunker to hide in. While the men fought to save the cottage, the women had to protect themselves, putting spot fires out with tins of water while struggling with stinging, smoke-filled eyes. The ferocity of the fire seared itself into Elizabeth's memory. Even after the passage of 40 years, her recollections bring to vivid life the characteristic features of fire: the noise, the wind and the heat.

> But the flames had no pity. They licked up with fury the grass, leaves and tender wildflowers, roared and crackled in the bushes, climbed the highest trees and seemed determined to devour all that was left. Oh what a scene it was—the burning heat and suffocating smoke mixed with gas from the burning gum trees caused excruciating pain to the eyes. What a babel of sounds—the roaring of the fire, the hoarse shouts of

> the men, neighing of horses and lowing of cattle. The wind rose higher. Pieces of flaming bark came whipping through the air. The flames impelled by the might of the mighty wind took great leaps and large trees were in a moment wreathed in fire to their topmost branches ... Night came, the country was illuminated for miles. Hundreds of trees stood like pillars of fire as far as we could see it was everywhere.[5]

In Gippsland the sky went as black as night in advance of the fire, although some reported that there was no smell of smoke to alert them as to the cause. The arrival of the fire soon clarified the origin of the darkness.

> Nothing can be conceived more sublimely terrible ... The flame careering with lightning speed along the tops of the trees, fanned and lashed on by the violent hot winds, is said to have been attended by the most appalling roar, more awfully overpowering than that of the ocean in storm.[6]

It must have seemed like fires were everywhere. From the west they raged across the Wimmera and in the south from Mount Gambier to Portland. The young city of Melbourne was encircled: from the farmlands around Geelong to Mount Macedon, from Seymour around into the Plenty Ranges, down into the Dandenongs and as far south as Dromana. The fires crept in as close as Diamond Creek and Heidelberg. Further east, fires burnt throughout the alpine areas, from Mansfield to Omeo and down into Gippsland, where the damp interior seemed to be the only large area of the new colony to have escaped unscathed.[7]

Yet despite over 5 million hectares—one-quarter of the

colony—being burnt on Black Thursday, the casualties were relatively low. Only twelve people were known to have died. Most deaths occurred in the Plenty Ranges, probably because of the density of small farms neatly dividing up the fertile valley. Here residents saw 'scores of well tilled farms and cheerful homesteads being changed in one short day into an area of charred desolation'.[8] Near Diamond Creek, the home of George McLelland was overtaken by fire. So sudden was the onslaught that his wife Brigid and five small children—John, James, Joseph, Mary Ann and William—were unable to make it even the short distance of 100 metres to the waterhole. George McLelland, badly burned, was the only survivor.[9]

Black Thursday 1851

There was a Black Thursday in Port Phillip, so called from the country being overwhelmed with fire and smoke, as if a destroying angel had winged its way through the air, scattering firebrands far and wide, its wake lit up with flaming forest, the fire and smoke, as if waging war with each other, spreading consternation and dismay throughout the length and breadth of the province. From an early hour in the morning a hot wind blew from the north-north-west. The atmosphere became so dense as to render outdoor life almost intolerable, for every mouthful of air was like a flame puffing out of a furnace. At 11 o'clock ... it was 117 degrees [Fahrenheit, 47°C] in the shade, at 1pm fell to 109 degrees [42°C]; but at 4pm went up to 113 degrees [45°C]. There was no electric telegraphic communication, so that it was not possible for the townspeople to obtain any intelligence from even a few miles in the country until next day.

It was not long before account of woe and desolation came trooping into town. East, west, north and south joined in the same refrain of the ravages caused by the bush conflagrations. Amongst the Plenty Ranges the calamity was hardly capable of description. The fire had, it was said originated in that quarter through the carelessness of two bullock-drivers who left some logs burning when they went away in the morning—these setting fire to the long drought-parched grass ... An unfortunate settler named McLelland lost his wife, five children, home and 1100 sheep. One hundred persons were left homeless and penniless. Every place was a scene of misery and lamentation; the dead carcasses of sheep, horses and cattle blocked up the waterways and thoroughfares; and an excursion such as I made in that quarter two days after was a sickening trip to take. Dr. A Thompson declared that not one house in ten was left on the Barrabool Hills, and all the small farmers were either burned out or ruined.

Mount Macedon was lit up in a style that would gladden the hearts of the Druids of antiquity ... and three men were known to have perished there. As to Kilmore, how it escaped was inexplicable as the country in every direction was a black, burned-up desert. No quarter of the district escaped, for the conflagration might be said to be general from Gippsland to the Murray, and from the Plenty to the Glenelg. At Mt Gambier, near the Glenelg, the township was almost completely extinguished. The Dandenong division was devastated by fire in such a manner that every vestige of tillage or verdure was burned from off the ground. On the Friday rain fell lightly in several parts of the interior.

Edmund Finn, 1888[10]

Thirty kilometres out to sea, fierce, hot winds buffeted the ships in Bass Strait. The nor'nor'westerly wind was so strong that one sudden gust ripped a ship's topsail from the bolt rope. Burning embers, ash and dust fell over the ship's deck.[11] The northerly winds blew the burnt remains of Victoria 600 kilometres across Bass Strait to the Tasmanian town of Launceston, where they darkened the skies and turned the sun the colour of polished brass.[12]

Only 12 human lives had been lost, but a million sheep and thousands of cattle had also died. While vast swathes of grassland and pasture had been lost, the impact on the ever-declining native forests was even greater. In this one fearsome season, 1.5 million hectares of forest was burnt out. This was not the mosaic patchwork of small fires the forests seem to have evolved with and adapted to. Many small fires had joined together into a megafire. It was a sustained, extensive conflagration across vast areas, making recovery for many species much more difficult.

Travelling through the Otways a few weeks after the fires, one visitor described the fierceness of the winds, which in the central forests had snapped huge trees, leaving an 'impassable labyrinth' of charred remains. On the coast, the fires had burnt so severely that the trunks had been stripped of their branches and the ground was burnt bare.

> The only vegetation was a microscopic fungus—blood-red. Bird, beast and reptile appear to have alike perished. Occasionally we lighted on the carcase of some one of the smaller animals. The smoke would doubtless have suffocated them had they escaped the flames.[13]

Some, however, felt that all Australians could learn from this tragedy. William Westgarth, a leading public figure in Melbourne, wrote hopefully in 1888:

> There has never been, throughout Australia, either before or since, such a day as Victoria's Black Thursday, and most likely or rather most certainly, it will never, to its frightful extent, occur again; for every year, with the spread of occupation, brings its step in the accumulation of protectives.[14]

Little did Westgarth realise that this was only the beginning. Australians have been taking lessons in bushfire tragedy, with frightening regularity, ever since. How much we have learnt from them remains to be seen.

Black Thursday was as much an environmental catastrophe as a human one. On the back of the widespread clearing, European farming practices seemed to bring with them an increasingly fierce and widespread regime of fire that would challenge the resilience of the men and women who had taken this fundamental feature of the Australian bush and unleashed it with renewed ferocity.

6 · FIRE IN THE OUTBACK

Fires in the outback don't often make it onto the national news or into the history books. But, as in most other areas of Australia, fires are an integral part of many grassland habitats and essential to the animals that inhabit them. Regular, cool burns prevent trees from growing and shading out the grasses. Fires are a key means of promoting grassland. A rich diversity of animals, particularly seed-eating birds, thrive on the grasslands of northern Australia. From the gaudy and colourful finches and parrots to the quiet and cryptic quails and pigeons, all depend upon the open savannahs for their survival. Without fires, the trees begin to encroach, along with the predatory butcherbirds that prey on young birds. Only in the thick grasses can brightly decorated finches find refuge.

Fire not only provides the structural habitat of the grassland birds but also regulates their seasonal food supply. For many species, the arrival of the wet season heralds the arrival of a time of plenty—but not for the seed-eaters. As the dormant seeds in the soil spring to life with an abundance of new shoots, they camouflage the seeds remaining in the soil, creating a sudden shortage of food. 'Storm fires' burn off patches of the new growth, providing foraging sites for the birds, particularly the vulnerable juveniles just fledged from the nest. Many of the grass species burnt at this time of year also go on to produce vastly more seeds later in the season.

Of course, humans like grass too, particularly to feed their sheep and cattle on. As humans moved into permanent settlements, they inevitably came into conflict with the very fires that probably maintain the desirable grasslands that attracted them in the first place. In country regions the loss of human life may be smaller, but the frequency of fires may be greater, and the loss of stock, livelihood and infrastructure can have profound effects across the community.

THE GARDEN CITY OF THE WEST

The new pool was the pride of Barcaldine. At the turn of the 20th century, the concrete rectangle was constructed 100 feet long and 25 feet wide, it was filled from top to bottom with nearly 60 000 gallons (275 000 litres) of the pristine waters from the Great Artesian Basin—water that also allowed Barcy's wide streets to be carpeted lilac in spring with the blooms of the jacaranda trees and caused the fences to sag under the scrambling weight of hot pink and purple bougainvilleas. The residents of Barcaldine prepared to swim, despite the Federation drought that had lasted for ten long years, dried the mighty Murray River to a trickle and left tens of thousands of sheep desiccated in the paddocks. In the dry heart of inland Queensland, Barcaldine was the Garden City of the West.

The inaugural Swimming Carnival was held in November 1908. Well-built young men in tight-fitting singlets and shorts jostled each other in front of the corrugated-iron dressing sheds, apparently oblivious to the ladies smiling and chattering behind hands and hats at the back. They were not the first swimmers to test the waters. That honour belonged to a staggering drunk who had had to be rescued from the deep end shortly after the pool's completion.[1]

Life was good in Barcaldine. Located at the end of the

railway line to Brisbane, Barcaldine was the transport hub for central western Queensland, steam engines feeding a steady stream of wool from the vast surrounding stations—Barcaldine Downs, Home Creek, Coreena, Avington, Delta, Evora and Saltern Creek—spreading out over a million hectares of spinifex and scrub. Eleven pubs lined the main street, filled with shearers, carriers, drovers and others—hardworking, hard-drinking men who'd come to build the railway and stayed.

Barcaldine was famous. Not just locally, mind you, but nationally. In 1891 the shearers had gone on strike. The union wanted better pay and a white Australia—no Chinese, Japanese, South Sea Islanders or other coloured races. The pastoralists brought in scab labour from the south. The strikers armed themselves for war, some accused of starting grass fires as they rode near Jericho into Clermont. Military reinforcements arrived. Barcaldine's population surged to 4500 people before the strike finally subsided. The rain broke the drought and strike at the same time. Muddy water turned the tent cities into quagmires and disease ran rampant through the striking workers. Funds from the south dried up as the heavens opened.

The fight was lost, but the Labor Party was born. Everyone knew that the home of the workers' party lay under the Tree of Knowledge in the main street of Barcaldine.

A town on fire

Barcaldine was built, not on wool, but on water. It was access to water that determined its location, and water that gave it a future. *Pugh's Almanac* enthused that Barcaldine

> is one of the prettiest laid-out inland towns of Queensland, possessing wide streets which are planted with magnificent trees, that in season are a blaze of flowers. Barcaldine, with its numerous bores of hot water and high medicinal and curative value, ought to be a sanitorium. It is veritably an oasis of the great western country.[2]

Barcaldine's water might have been pure, healthy and abundant in 1908, but, without a pumping station and reliable delivery system, it flowed at a trickle.

In the early hours of 10 August 1909, a fire started in Mr Dias's store on Oak Street. With no water to stop it, the fire roared through the neighbouring timber buildings, consuming 18 shops in two hours, setting off firecrackers in Wah Sung's shop and destroying the Commercial Hotel. Demolition, not water, was the only thing that stopped the fire spreading further.

> Daylight revealed a sorry scene—smoking ruins, piles of rescued belongings, and partly pulled down buildings. Along the school fence were pianos, sewing machines, buggies, horses, birds in cages, bedding and furniture. Atop it all Mrs. Kemp's cockatoo was reported to be whistling, 'There's no luck about the house.'[3]

The disaster led to the formation of a volunteer fire brigade that year—two teams of men were provided with hoses, hydrants and a pump on iron wheels known as Gentle Annie, to be joined in later years by Red Wings. They still had no water but they were keen, and eventually a 200 000-litre tank was built atop a 30-metre tower, providing reliable water for the townsfolk and firefighters alike.

WAR

In August 1914, news trickled into Barcaldine that Britain, on behalf of the empire, had declared war against Germany. According to the soon-to-be prime minister, Andrew Fisher, Australia was ready to 'stand beside her own to help and defend her [the Mother Country] to the last man and the last shilling'.[4] Barcaldine greeted the news with loyal dedication. A convoy of 'sturdy, bronzed, robust young fellows'[5] (around 300 in all) waved farewell to their families as their train pulled out of Barcaldine Station. Some of their sisters signed up as nurses.

Recruits weren't the only ones to leave town. Germans were held in great suspicion. Chinese shopkeepers and market gardeners had to be registered as 'aliens'. Even Swedish Anne Enemen and some Japanese men were registered. Many such aliens left the country at the beginning of the war. Some never returned. And as the young men disappeared, work on the stations had to be taken up by their fathers and uncles, their mothers and grandmothers.

The British government purchased all the wool Australia could produce for a fixed price for the duration of the war. Uniforms, blankets and socks were in short supply for soldiers on the icy Western front. Wool was an essential industry.

SALTERN CREEK STATION

Despite the war, life on stations like Saltern Creek continued much as it had since the station had first been claimed in 1872. The last few seasons had been good. The drought had broken in 1916 with 66 millimetres of rain, followed by 89 millimetres the following year. January 1918 brought floods across inland

Queensland. March brought more heavy storms. By April they'd had 58 millimetres; ample to see them through the Dry. The grass was long and thick, providing plenty of fodder for the cattle and sheep. Sheep numbers had almost returned to their pre-drought levels. A careful program of breeding saw a steady improvement in the wool quality, slowly building up the reputation of the inland stations for fine-grade wool. Mollycoddled champion rams were put with the best ewes, each generation of lambs an incremental improvement on the last. They finished the shearing in the cool, dry winter months. Saltern Creek fleeces sold steadily at two shillings and fourpence in London. Barcaldine, like Australia, rode comfortably on the sheep's back.

The cluster of houses at the heart of Saltern Creek Station was a hive of activity, from the butcher's shed to the blacksmith's forge, from the well-stocked store to the little weatherboard workers' cottages. The long shearing sheds may have been empty but the station was still busy with daily life, a little township in its own right. It was an hour's ride just to get to the main road, another few hours to get to Barcaldine. At the heart of this empire was the old homestead, with its long, wide, thatched verandah shading the timber structure on all sides. The neat lawn was bordered by flower gardens, cannas producing a riot of colour interspersed with white gravel paths.

The first rumble of thunder heralded the end of the dry season and the beginning of the Wet. The cool nights had started to warm, the clear skies giving way to clouds. But the rain had not yet arrived. As the dry inland air heated up and moved east, it hit a stream of moist, unstable, coastal air. Thunderstorms accumulated at the collision point.

Lightning cracked across the sky, illuminating the paddocks. The cattle bellowed in fear, stampeding blindly across the open plain, running against one another, their eyes rolling white. Soon the sky was alive with lightning, fracturing the purple sky with dizzying nets of energy. Fingers stretched out, blasting trees in a shower of sparks and embers. The fires spread down the trunks, into the long grass, growing like a black fungus around their bases.

Lightning regularly started 'storm-fires' at the end of the Dry. These quick, light grass fires spread rapidly and posed little threat to human life. During droughts there was little enough grass to burn anyway, and the flames might be low enough for a man or horse to step over. But after a good wet season, the grass was thick and luxuriant, high enough to lose cattle and sheep, too tall for a man to see over. Some farmers burnt the old grass on their properties before the storm season, reducing the fuel load in risky areas like gullies and along waterways. But not all farmers did this, perhaps unwilling to lose pasture before the new grass emerged. The very feed that fattened the calves and lambs and ensured prosperity carried in it the seeds of its own destruction.

Since September 1918, bad fires had been burning across inland Queensland. Some could not be contained, despite the best efforts of local men to stop the fronts reaching the foothills of the ranges. Once the fire got into the hills, nothing could stop it. On an uphill slope, fires double their speed as the rising air preheats the grass ahead. Even slow, downhill fires offered little relief as the damper gullies and ravines were thick with vegetation and fuel. Fires burnt across the region, from Charleville to Blackall, Barcaldine and Hughenden.[6]

At Saltern Creek station, the manager, Arthur Jones, marshalled his troops. He had battled flames before and knew what to do. He had good men at his disposal, like Edward, an Aboriginal station hand who had a cool head and could keep men on task in the face of the onslaught. The town might be protected by a fire brigade but the surrounding stations defended themselves. Fires make firefighters of all station hands. Managers become fire captains, deploying troops and directing resources, carefully weighing up risks to life and property. Experienced hands lead teams of beaters to dampen the flames and construct crude fire breaks with rake hoes intended for gardening and sowing seed. Men were brought from far and wide in motor cars, even a double-decker bus, to help.[7] They used whatever equipment came to hand—buckets, sacks, hoes, even branches from trees.

But the fire of Tuesday, 29 October 1918, was not like the other fires Arthur Jones and his men had fought. The grass was long and thick. Lightning and embers lit new fires faster than men could be deployed to extinguish them. Twenty different fires were burning across Saltern Creek Station. Not even the British Army could have put them all out, according to one observer. The flames burnt higher and more fiercely with more fuel, and travelled faster than they had in the past. The station men, never complacent about the dangers, were taken by surprise when the wind changed.

Little lines of fire raced across the paddocks, low enough for a man to jump over. When the flames reached the thicker grass and bushes of the gullies, they roared into life, heat searing the surrounding air, burning lungs and singeing flesh. The change in wind direction brought not relief but disaster. The fire flank at Willoughby Station erupted, surging forward and trapping Arthur Burgess and young Alfred Hock against a fence. Severely burnt, the men were rushed to the Barcaldine Hospital, where they died within hours of each other.

Arthur Jones and Harry Knowlton succumbed as their co-workers tried to drag them to safety, their clothes burnt from their bodies. Nearby the wagonette was charred around the wheels, but the paper wrapping from their bread fluttered about, unburnt, suggestive of safety just out of reach. Harry Gilchrist, Roy Holmes and Blaney Thorpe were also taken to hospital where they later died. On nearby Rodney Downs, the station manager and his bookkeeper only escaped the flames by galloping through them.[8]

The loss of human life was a tragedy but the impact on the broader community was immense. Stock losses, lack of feed, loss of fencing and infrastructure profoundly affect the viability of properties. Improvements built up over generations were lost within a few days. Without stock, without feed, there is no prospect of income for the coming season, just greater expense as the damage is repaired.

Across the vast acreages lay over 100 000 sheep, incinerated by the flames. The cattle fared slightly better, often able to run through the flames to the safety of burnt ground. Kilometres of fencing was reduced to regular piles of charcoal and ash. Stock not yet dead wandered silent across the blackened landscape, slowly starving to death, unable to eat or drink through burnt mouths and throats. The surviving herds merged and mingled across the stations, seeking the solace of their own kind. The grim task of culling began. Downers were dealt with first, those to weak to rise. The animals that had taken refuge in the creeks and dams had to be hauled out, exhausted and depleted after their ordeal. The men gathered the survivors into yards, culling those too badly burnt to survive, those with swollen limbs, cracked hooves or laboured breathing. They would not live. The air was sickly sweet with burnt and rotting flesh.

The tragedy spans time and space, farming communities united by their losses. On a farm on the Nagambi Ranges in the ACT, nearly a century later, a farmer could have been describing the same scene as at Saltern Creek.

> Hundreds of sheep were trapped in mud on the edge of the dams. They were plucked out, as they would have died had they stayed there. Most were motionless, in shock. Many lambs were dead or dying, having breathed superheated air and scorched their airways. There was no hay or fodder for feed ... some were unable to eat grain; they died of starvation.[9]

Late in November, the school, fire and church bells rang again across the town of Barcaldine. Men leapt from their beds ready to do battle again but this time the battle was over. Not the one so recently fought on their own lands but the one fought on far-distant shores. The war had ended. 'Peace! Peace!' went the cry as crowds gathered on the streets clanging kerosene tins and cow bells. Relief swept across the town, too tired to celebrate, still mourning its losses both here and abroad.[10]

Thirty-eight men from the Barcaldine district lost their lives in World War I. Seven men died fighting the fires on the stations. There are memorials to them all in the district. As bushfire researcher Neil Burrows says, 'The people that volunteer to fight bushfires, it's as dangerous in many ways as volunteering to fight a war.'[11]

EYRE PENINSULA 2005

Port Lincoln shelters on the tip of the Eyre Peninsula, an isolated oasis bounded by green national parks, the Southern Ocean and a great expanse of desert. Low hills surround a capacious harbour which is home to a thriving fishing fleet. The giant, white wheat silos glow in the sunlight like an ancient Colossus of Rhodes, a symbol of fertility and abundance on the edge of a dry land.

Inland from the harbour, beyond the close-cropped hills, the land flattens out into golden wheat fields interspersed with pockets of dark green scrub. Fires are an integral part of the Eyre Peninsula landscape. Frequent grass fires burn quickly but relatively coolly across the open paddocks of the peninsula. They spread rapidly but do relatively little damage. The danger lies where the fires get into the tea-tree scrub and low forest that dot the peninsula. Here, in thick, low vegetation, the fires burn hot and fierce, creating an ember bed for further fire spread.

Mallee

The vegetation on the Eyre Peninsula is dominated by mallee—the name referring not to a species of eucalypt, but rather to its growth habit. In poor soils with low rainfall, many eucalypts become mallees, characterised by a low, multi-stemmed coppiced trunk. These stems reshoot from a massive mallee root—the underground lignotuber used by so many eucalypt species. Inured to the hardships of their environment, these stunted woodlands respond to fire with enthusiasm, along with the birds, beetles, reptiles and mammals that make their homes here. Few species here are sensitive to fire and many respond positively to fire, moving in to freshly burnt area to take advantage of the clear ground and exposed food. Cool frequent fires maintain a patchwork of diversity of species and the rich mix of mallee woodlands and open grasslands extends across the southern region of South Australia and across the long, flat plains of western Victoria.

As a child I lived on the edge of this bush, looking north across the wheat fields. We lived in a caravan, near a cluster of sheds, at the end of a long, sandy track that wound its way through wire fences and scrub. In winter, the water froze solid in the tap from the rainwater tank and we shivered as we showered under a leaky bucket in one corner of the shed. In summer, the brown snakes emerged to sun themselves along the edge of the driveway. The grass cured gold and the broad paddocks shimmered in the summer heat. The water in the tank fell, the scarce resource attracting flocks of galahs racketing across the empty skies. Sand scorched across the desert, stinging bare legs and burning eyes.

I remember my parents' concern as they looked north, across the dusty fields, where dark smoke regularly smudged the horizon in the distance. I recall anxious conversations about the dangers of the long, windy drive and the flimsy caravan, the long grass and the dense bush. I remember the unfamiliar glow to the north dimming the bright stars on hot nights. I looked doubtfully into the murky green slime of the concrete tank, with its ominous collection of dead galahs and wondered, as children do, if I would end up in there too.

Before summer's end, we had moved to the carefully manicured, perpetually green lawns of the local caravan park, cooled by sea breezes blowing across the bay. At the ocean's edge, this oasis seemed secure, safe from the untamed dangers of the dry inland. We felt safe from fires here, just as the residents of nearby North Shields felt safe, so close to the beach.

Coast and country are almost different worlds on the Eyre Peninsula, just as they are for most of Australia. We cling to the periphery of our continent looking out over the blue seas, and a life of fishing, beaches and holidays. Inland lie the harsher truths of dust and drought, farms, sweat and flies. In the coastal townships of the Eyre Peninsula, there is less awareness of the risk of fires. Inland, the fires are an integral part of every summer, a risk to be planned for, responded to and recovered from.

The pattern of fatalities in the Eyre Peninsula fires reflects that of many modern bushfires. Two of the deaths were men engaged in firefighting activities. Six were women and children who died in cars attempting to flee the flames. The final victim was a teacher at the local high school who, many years ago, first inspired my interest in studying biology. Helen Castle died at North Shields sheltering in her bathroom. Firefighting outside, leaving late and passive sheltering are historically the three riskiest activities of all in a fire.

---

10 January 2005

Late Monday afternoon a white Toyota Landcruiser left a gathering cloud of dust in the distance as it trawled along the dirt track, metal detector bouncing in the back. Spying a promising pile of rocks, the driver pulled over into the long grass. Cicadas vibrated the hot air around him as he stalked across the open paddock. The sheoaks whispered softly in the growing breeze. The fossicker wiped his forehead and returned to his car. He sat in the air-conditioned cool of the cab with the engine idling, and knocked back a soft drink.

Putting the car into gear, he pulled back out onto the road and headed for Wangarry. As the engine revved, a burst of heat swept down the exhaust system, dislodging a shower of red-hot embers out of the muffler and into the bed of dry grass. The car disappeared into the dust. Within twenty minutes, the South Australian Country Fire Service had been alerted to a fire in the area. Tankers and farmers converged on the grass fire, anxious to extinguish it before it grew too big. Tractors, graders and front-end loaders joined in, carving barriers of bare earth in an attempt to contain the fire's spread. By late evening, large areas of land were being burnt ahead of the fire, starving it of fuel in anticipation of its arrival. By nightfall, the fire was deemed contained. It was still burning but within a defined area. Most of the fire crews returned to their stations; only blackened paddocks and a few smouldering logs remained.

The farmers and firefighters went home, tired but wary. The weather forecast was bad. A fire tomorrow could be much worse. In town, office workers and residents watched footage of the fire on the news, sandwiched between reports of the deteriorating conditions in Darfur and a bus catastrophe in India. The weather forecast was bad. At least work was air conditioned.

Tuesday dawned hot, dry and windy, just as it does on many summer days on the Eyre Peninsula. People went about their business, driving to work through the suburbs of Port Lincoln and walking quickly through the shopping strip and business district, anxious to get out of the fierce sun. It was going to be a stinker.

By ten o'clock, several spot fires had flared out of the old fire ground, fanning out to the south-east, into Wanilla Forest. Local residents rushed to defend their own and neighbouring properties, calling friends and family. Some leapt into action, knowing what to do. Others waited, unsure of the right course of action. Alerts went out on the radio, for those who were listening. Fire trucks arrived and left in a hurry, responding to crackling messages on the two-way—no time to talk today. The fire stretched black fingers across the peninsula, hot fierce tips ringed with a smouldering perimeter along their length.

Just before lunch the wind changed to the west. The long smouldering edges of the fire roared to life with new fuel and oxygen. The winds drove the widened front across the peninsula, over farms and houses, towards the coast. The fire fled across the grasslands, faster than anyone expected, consuming the paddocks and shrubs in its path. Under the dark, smoke-filled sky, no-one could see where the fire was, where it was going. Flames seemed to come from nowhere, suddenly bearing down on a house, a road, a car. Confused, families drove their cars towards flames, instead of away, leaving behind the safety of houses that would still be standing after the fire.

Along the coast, little pockets of development radiate out around Boston Bay. At North Shields, along the Lincoln Highway, a collection of fibro houses and caravans cluster around the airport and the beach, protected from the dry inland by a ridge of low hills. At lunchtime, black clouds billowed over the hills, sending ripples of concern through the small community. Flames licked the summits, before spreading downhill across the paddocks and into the township, before finally burning themselves out at the ocean's edge. Unprepared for the onslaught, residents fled in cars and on foot, some to the airport, others to the beach. Others stayed at home as the flames erupted around them.

By nightfall, nine people had died and nearly 80 000 hectares of land had been burnt. Countless homes, farms and lives were left shattered in the fire's wake.[12]

THE BIGGEST BUSHFIRES EVER

Natural disaster and the year 1974 are inextricably linked in Australian memory. The day before Christmas, Cyclone Tracy destroyed Darwin, flattening the city and killing 71 people. In a territory of just 100 000 people. Cyclone Tracy was devastating but it was not the only natural disaster that the Northern Territory faced that year. Over the summer of 1974–75, the Northern Territory was seared by the biggest bushfires ever recorded.

It was one of the worst fire seasons ever seen in Australia. The country had been in drought for the past three decades before exceptional rainfall produced a proliferation of grass growth in central Australia. More than 460 millimetres of rain had fallen at Finke in 1974, more than double the average. As the temperatures rose over the summer, the luxuriant grass dried in the heat. In themselves, these conditions were not particularly abnormal. Then in January, dry lightning storms swept across inland Australia, setting the desert on fire.[13]

Fires raged across the arid and semi-arid heart of the country. From Kalgoorlie in the west to Urandangie in the east, from northern New South Wales and South Australia and all the way north. Six people died in this fire in the far west of New South Wales, where stock and farming losses were also high. Even the scale of individual fires was astonishing. The Moolah–Corinya fire, with a perimeter of over 1000 kilometres, was surprisingly extinguished by firefighters rather than by the weather.

It was the Northern Territory that fared the worst. Almost one-third of the entire territory burnt that season. Across the country over 117 million hectares—15 per cent of the entire continent—was burnt in a single season. That is more than five times the size of the state of Victoria, almost five times the size of the United Kingdom. It is an area 23 times the size of the 1851 fires. No other fires come close to it in scale. In the United States of America, the largest area of land burnt in a single year is just 3 million hectares by comparison.[14] And yet this fire season is all but forgotten.

The Great Chicago fire of 1871 killed 250 people and has been immortalised in American, if not world, history. Yet on that very same day, 8 October 1871, between 1000 and 2000 people died in wildfires in rural Wisconsin. No-one knows the exact scale of the tragedy that wiped out the frontier town of Peshtigo and the surrounding 600 000 hectares. Many residents were itinerant workers and native Americans. Records of who lived in the area at the time were poor. Many bodies were buried unidentified. What is known is that the Peshtigo fire remains the worst bushfire tragedy in history. Yet unlike the famous Chicago fire, the Peshtigo fire for many years all but faded into the mists of local history. Why?

We could argue that the inland Australian fires of 1974–75 are forgotten because they occurred in areas with few people, with few casualties. Despite their massive extent, these were mild fires posing little threat to people and property. The same cannot be said of the Peshtigo fire. Rural disasters just don't seem to evoke the same horror that urban disasters do. Perhaps it is because most people live in cities that urban disasters frighten us more—making it more likely these disasters could happen to us. Maybe it is because we expect nature to be a little wilder, a little less controllable, in the country. The city is a human creation that seals us off from nature, protects us and shelters us. When nature breaks down that barrier, we realise how truly vulnerable we are. James Mitchell argues that 'when earthquakes, floods and storms strike cities they intrude on places that are monuments to the human control of Nature'.[15] At those moments, we realise just how tenuous that control really is.

The fires of 1974–75 are all but forgotten because they occurred in areas where few people live. Arguably, their very scale is a feature of their isolation—there is no reason to put out a fire that threatens no-one. Grass fires like these burn fast but cool. They burn in areas with an abundance of cleared land where they are comparatively easy to avoid, not in the densely vegetated, heavily populated areas of the southern districts where the proximity of people and trees makes for a volatile combination.

It is easy to argue that these fires are all but forgotten because they were not as severe or fierce as the southern forest fires. But their sheer scale and scope defies this argument. Should we let the biggest bushfires in the history of the world pass unremembered? Is there nothing we can learn from these fires?

The forgotten story of the 1974–75 fires show us that fires—enormous fires—are an integral, almost unremarkable, part of the Australian landscape. They show us that even huge fires do not always have to have huge death tolls—that it is people, not fires, that make for tragedy. Our risk lies not so much in the nature of our landscape but in how we chose to live in it.

Surely this is something worth remembering.

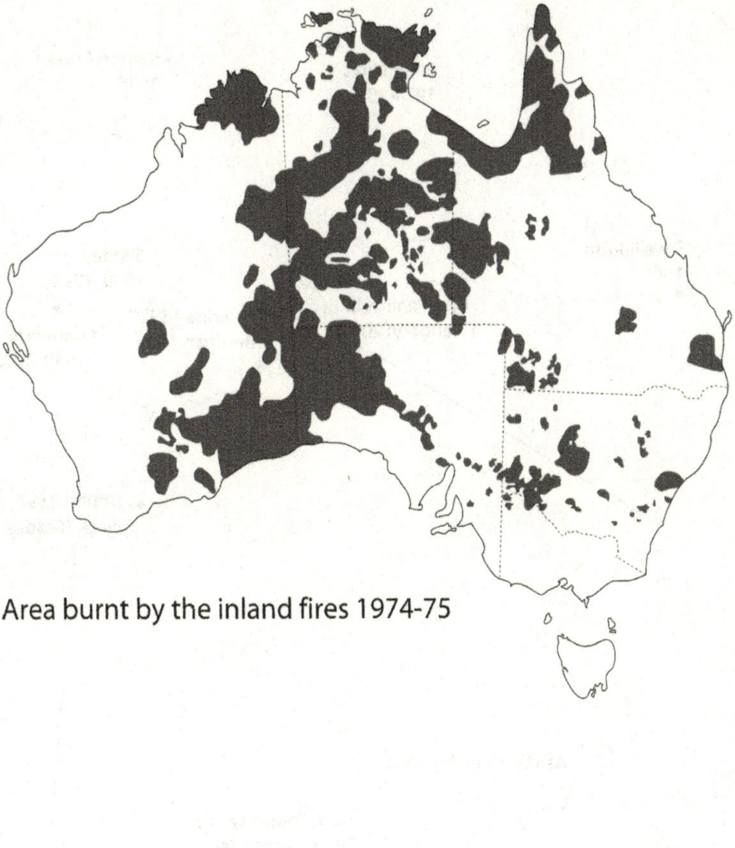

Area burnt by the inland fires 1974-75

Locations of some of Australia's worst fires

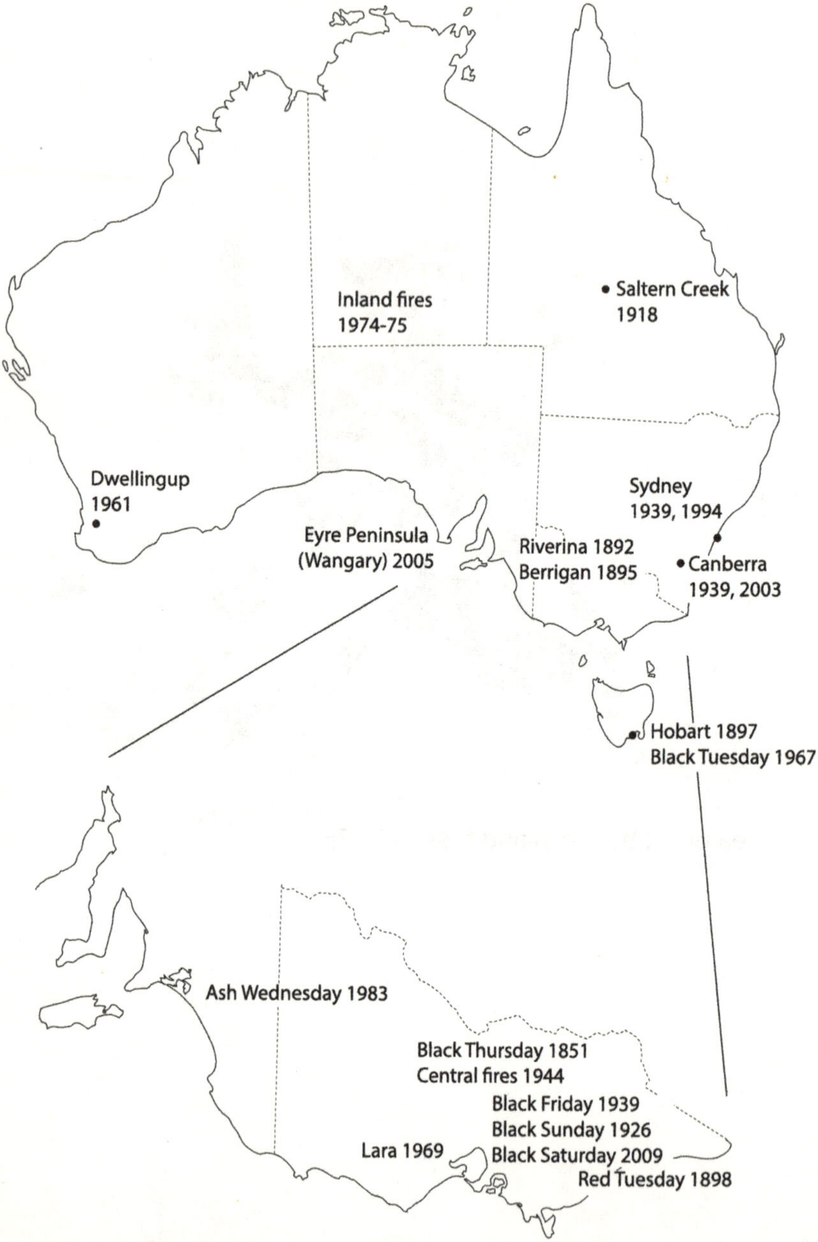

7 · LIVING IN THE FOREST

Fire in the forest. The bushfire—scourge of the Australian summer, sweeps paddocks and wheatfields in all states.

But in Victoria, where much of the country is densely timbered, the flame has become a menace not only to property but to life. More than a score of people have perished already in fires that reach over a vast area of Victoria, wiping out timber camps, townships, sawmills and hydroelectric stations.

Gallant volunteers fight desperately to save the homes of lonely forest dwellers but their efforts are mostly in vain. Four persons fleeing in this car were trapped when the car broke down. And in this car three more were burnt to death.

Night brings no mercy from the agony of fire. Hundreds cut off by the flames are reported to be hiding in dugouts, hoping the peril of heat will pass them by.

Once fire gets loose in these forests, dried by months of drought, all the efforts of men seem futile.

Yet men fight on courageous, resolute, giving their strength, risking their lives in the hope of saving life and property from the mad wrath of fire.

Movietone News, 1939[1]

The western and eastern tips of the Australian continent reflect each other like images in a distorted mirror across the great expanse of the Bight and the Nullarbor. They are opposites—the same yet different, each containing paired species that resemble each other yet are subtly different. Where western quolls, or chuditch, stalk the tall forests of the south-west, the eastern spot-tailed quoll inhabits the forests of the east. Western grey kangaroos dominate the plains in one direction; eastern greys monopolise the east coast. The western ringtail possum finds its home in the tall forests of marri and jarrah; the common ringtail clings to the eastern forests of ash and stringybark.

Despite their profound differences, the forest of the south-west and south-east share one thing in common: a frightening propensity to burn. Their history unites their separate evolution with a common theme.

BLACK FRIDAY 1939

In 1939 Australia was slowly recovering from the Great Depression, which had seen widespread poverty, hardship and unemployment across the country. In the final year of a difficult decade, Australians might have looked forward to better times ahead. But, as dawn broke over a hot, dry continent, the new year did not look so promising. Fires were dotted across the southern and eastern states: some started by lightning, some by camp-fires, many by sheer carelessness. Most were in isolated areas, burning through tracts of inaccessible forest and mountains. No-one gave them too much thought. If the forestry workers in the mountains were worried, they just shrugged and went back to work. There wasn't much you could do about fires.

As January progressed, the weather deteriorated, getting progressively hotter, drier and windier. Fires flared up in Tasmania, threatening the prime minister's house at

Devonport. On 11 January, fire swept through the Adelaide Hills, burning out the towns of Macclesfield, Mount Torrens and Gumeracha. The following weekend, fires raced down from the mountains towards the teenage city of Canberra, burning almost one-third of the ACT. The fires burnt into pine plantations and threatened the Mount Stromlo Observatory, causing consternation over Canberra's poor fire readiness and slow response. One person and more than 2000 head of livestock died.

Around Sydney, the situation was just as bad. On Saturday, 14 January, fires ringed the city in the areas of Sutherland, Warringah and Baulkham Hills. Westerly winds drove burning embers into the city itself, driving the fire from Windsor to Rouse Hill and Galston, just west of the Ku-ring-gai Chase National Park in Sydney's north. Further north, the historic town of Seaham was almost completely destroyed by the fires. The smoke could be seen as far away as New Zealand, which suffered in the same heat and had its own fierce fires to contend with in the Hawkes Bay area. Nor did night bring relief, with temperatures dropping to a minimum of 34°C and rising the next day to 46°C. Thirteen people died in the fires across New South Wales and over 70 000 hectares of land burnt.[2] But this tragedy was all but overshadowed by the magnitude of the calamity that had occurred in Victoria.

The fires in Victoria had started early after a dry year. Winter had barely ended when the first serious outbreak occurred, a crown fire at Creswick. The spring rains largely failed and, in October 1938, fires broke out across the western highlands and Gippsland. Then, in January, the real trouble started. In a characteristic summer weather pattern, a large high-pressure system squatted over the eastern half of the continent, pushing cooler fronts to the south and stoking an enormous stockpile of hot air over the land. The south-eastern corner of the continent fried in low humidity and temperatures over 45°C.

Fire had already claimed the lives of two district foresters, Charles Denby and John Barling, on Sunday, 8 January. Twelve more died at sawmills within the burning forests on Tuesday. By Friday, 13 January, the small forest communities were already on a war footing, attempting to bury their dead while still fighting and evading an ever-strengthening enemy. The strong northerly winds fanned the fires, which in turn created fire-breathing, gale-force windstorms that roared through the forests. The pall of smoke that had hung over townships for days suddenly heralded their imminent destruction. Residents fled to creeks, makeshift dugouts, rainwater tanks, sawdust heaps, mine shafts and tunnels. Many died—drowned, boiled, suffocated, crushed or simply incinerated.

As the cool winds brought respite for the fighters in the early hours of Saturday morning, a solitary figure emerged from the melted remnants of a sawmill in the Matlock forest. Silent among the forest of cenotaphs, George Sellars, ashen faced and clutching his blanket, was the sole survivor of 16 men at Fitzpatrick's Mill. There had been no dugout or fire protection facilities at the mill. The forest was not considered high risk. No-one had thought it would burn with such ferocity.[3]

Over the season, nearly 1.5 million hectares, more than a quarter of the entire state, was burnt. Most of those fires were ferocious forest fires, burning through the state's inaccessible mountain regions from the outskirts of Melbourne right up to the banks of the Murray River. The ecological damage wrought on the Victorian landscape, not least the magnificent mountain ash forests, home to the tallest hardwood trees in the world, may never be repaired.[4]

The emotional damage to those who survived stayed with

them for the rest of their lives. The social repercussions of the fire, manifest in a landmark Royal Commission into the tragedy, remain with us today. The forestry departments and services were united under the *Forests Act* and given greater responsibility for fire protection. The Country Fire Authority was eventually formed in 1944, uniting disparate fire brigades. Sawmills were no longer to be built in the forests, but were located in nearby towns, where workers and infrastructure would be safer from fires.[5]

The politicians gritted their teeth, determined that this time would be the last time Australia would face such a fire catastrophe. From Canberra, the Minister for the Interior (and later Prime Minister) John McEwen declared that if we did not learn from the 1939 fires, if we did not take real action to solve this problem, 'Australia may as well reconcile itself to the sad fact that nothing will ever be done'.[6]

Hot spot

The south-west tip of Western Australia lies far from the monsoon rains that dominate the north, and beyond the desert country that spreads inland across dry clay-pans, sandy dunes and gibber plains. The sparse vegetation of the arid regions dots the landscape rather than covering it. There is too little fuel to generate a decent fire. The spaces between the plants prevent small fires from spreading.

But as we travel south down the west coast, Southern Ocean air streams in from the south-west, moistening and cooling the climate. With rising rainfall, we move from a landscape dominated by a few species to a landscape crowded with a multitude of species. It is fire that creates this diversity. The heathlands of Western Australia are renowned for their beauty, particularly in late winter and early spring, when they erupt with a riot of colours and blooms. This is one of the planet's biodiversity hot spots. Not only does it have a great many different species, but a great many of these species are found nowhere else in the world. Fire acts like a dispassionate predator, exerting its selective pressure without fear or favour. Without fire, more competitive species would ultimately exclude weaker species. Fire prevents any one species from dominating, allowing a much wider variety of plants to flourish.

The West Australian heathlands have their share of fire-dependent plants. The broad-leafed brown pea requires fire to germinate its seeds. The nitrogen-fixing fungus in its roots is the primary diet of the brush-tailed bettong, which also shelters in the shrub. Spores from the fungi are dispersed by the brush-tailed bettong in its faeces, which are also distributed by scarab beetles. Like the eastern barred bandicoots in Tasmania, brush-tailed bettongs thrive on the fungus, which is often exposed after fire, illustrating the complex interrelationships between plant and animal, fire and fungi.[7]

DWELLINGUP 1961

In 1961, Dwellingup was a lovely place to live. Perched on the top of an escarpment, it was surrounded by statuesque forests of jarrah, blackbutt and red gum, with underlying grass trees and banksias, paints and ferns. Just a couple of hours' drive from Perth, Dwellingup had most things you needed: a draper's shop, post office, fire station, school, hospital, two petrol stations and a hotel. The heart of the town was at the regional forestry headquarters—timber was Dwellingup's reason for existence, its main source of employment and even the basis for its social structure.[8]

Dwellingup, with a population of 300, was a neat little town of weatherboard houses and green lawns enclosed in white timber fences, shaded by the surrounding forest. The blokes went to the pub on Friday night for a game of darts and a beer while the girls got together in each other's homes with tea and cake. A generator provided electricity to the town between the hours of six and nine in the evening, but that didn't stop the occasional all-night party.

In the middle of January 1961, Perth was in the grip of a heat-wave. Most in the city had fled to the beach. The winter had been wet, promoting thick, luxuriant growth in the forests, which now shrivelled in the heat. Cyclones in the Pilbara sent strong winds and thunderstorms south, having already dropped their rain in the north.

On the night of Thursday, 19 January 1961, lightning strikes started 10 fires in the forest around Dwellingup. The forestry workers were put on alert, many called out of their beds to fight the fires in the forests. Firefighting was part of their job, an integral part of summer work. Their wives kept the woodstoves burning, ready to provide a hot meal when and if they should return. Fires are usually easier to control at night, when conditions are cooler and winds often drop. But this night was different. Temperatures stayed in the high 30s and strong winds gave the fires fresh breath.

On Friday night, lightning started another nine fires. By Saturday most were controlled, but only just. A strong northerly blew the fires to within 6 kilometres of the town. Under the dangerous conditions, firefighters were withdrawn from the forests.

Despite many fires burning across the south-west of the state, reinforcements arrived on Sunday to help the foresters at Dwellingup. A huge mushroom cloud hung over the town. At night, the hills seemed covered with fireworks. From Perth you could see that someone in the forest was really copping it.

Monday was washday in Dwellingup. Fires or no fires, life continued. Tomorrow the generator would run for an extra two hours to allow the ironing to be finished. Too bad if the washing hadn't dried yet. The men spent their fourth night controlling the fires, holding the line to the north of the town. But the weather was against them. Another cyclone in the north drove more fierce winds south.

On Tuesday, 24 January, the winds picked up. The nearby forest towns of Holyoake and Nanga Brook had been evacuated and Dwellingup's population had doubled. The fires to the north joined up into a single front and swept south with the sound of a thousand trains. As the fires intensified, radio contact with the firefighters was lost. They were all were ordered to retreat to Dwellingup, but no-one could be sure who had heard the message. The roads around the town were cut off and buildings on the outskirts started to burn. Families rushed to the hotel, the school and the oval to shelter. Embers set the children's hair on fire as their mothers covered them with towels. A few tried to get out of town, driving into a wall of fire instead.

All the buildings in the main street were burning and drums of fuel at the petrol station were exploding. House after house went up in flames. By the time the fire had moved through, the town was totally destroyed. Ninety people were missing or unaccounted for. The forestry department officials feared the worst. It had been their job to protect the town, to put out the fires, just as they had done for season after season before. Many of them felt they had failed.

Dwellingup was destroyed by the fire, along with Holyoake and Nanga Brook. Among the twisted wreckage of many houses, all that could be retrieved was the occasional kettle or an old-fashioned iron. Men lay in rows on the grass under the shade of trees, being treated for eye damage and smoke inhalation. But no-one died. The foresters may have lost the town, but their high level of organisation and the decisions to evacuate vulnerable towns and move out of the forests to the safety of the oval undoubtedly saved lives.

A Royal Commission into the fires was a turning point in forest fire practices in Western Australia. Regular prescribed burning to reduce fuel loads became routine, fire prevention technologies were upgraded and an active program of research into fire behaviour began. Dwellingup was rebuilt. But most of the surrounding timber settlements, like Holyoake, Nanga Brook, Marrinup and Banksiadale, disappeared. Life in the forest changed forever.[9]

In 2007, fires again surrounded Dwellingup as firefighters battled to save the town. Residents evacuated or prepared to defend their homes. Sixteen houses were lost. Some things about life in the forest never change.

8 · URBAN IMPACT

Fire is a fickle servant. It warms, lights, cooks and protects, but it quickly turns to bite the hand that feeds it. You would think that out in the open, living in the forests and plains, people would be particularly vulnerable to fires. In fact, it is when we crowd together in little, flammable houses, each with their own flickering flames of light and warmth, that we are most vulnerable. The history of human vulnerability to fire is a history of population growth, expansion and urbanisation.

Humans first started living together in large communities 3000 to 4000 years ago. Over time, these centres of economic activity expanded from small communities to towns to small cities. But for many centuries, the largest cities, the centres of Egypt and Babylon, were home to fewer than 100 000 souls. It was not until the start of the Common Era and the fall of the Roman Empire that the cities really began to expand, growing rapidly to populations of up to half a million.

With this expansion came risk. In the often narrow and crowded streets, the very buildings that protected and sheltered became pyres that trapped their inhabitants in great conflagrations. In cities where most of the buildings were made of wood, accidental fires were common and spread rapidly. The histories of many cities are marked by such tragedies. The cities with the most people suffer the greatest tragedies. The city of Rome famously burnt under Nero, although this was far from being the only fire it suffered. The Great Fire of London in 1666 and the Chicago Fire of 1871 have passed into legend. For all its infamy, only six people were known to have died in the Great Fire of London. The death toll for the Chicago fire numbered in the hundreds, a fairly typical toll for the fires that increasingly ravaged the cities of the 17th and 18th centuries. Rarely did the deaths extend into the thousands.

These famous fires pale into insignificance beside the great fires of Tokyo from the Edo period of 1603–1868.[1] The combined effects of earthquakes, lanterns and paper-and-timber houses resulted in devastating city fires that consumed whole suburbs and quarters, leaving a cleared and desolate landscape marked only by the characteristic stone lock-ups in which the Japanese placed their most precious belongings. At least 89 fires raged through the city during this time—one every three years. Wooden buildings were said to have a lifespan of just seven years before they would be burnt down. Tens of thousands of people died in these fires. In the worst tragedy, the Meireki fire of 1657, more than 100 000 people are recorded as having perished.

Cities burn from the inside out. They are burnt by structure fires—a burning building that spreads from one structure to another, slowly igniting whole neighbourhoods. As cities became better planned and less flammable, as concrete replaced timber, as fire brigades became more organised and water supplies more reliable, cities became less vulnerable to fire.

But in Australia, the risk for cities came, not from the inside, but from the outside. Australian cities did not burn from the inside out, but from the outside in. As Australian cities spread their green and leafy garden suburbs out into the fire-prone Australian bush, the potential for disaster grew.

HOBART 1967

Karen Schofield was just one of many children coming home from school that Tuesday. It was 7 February 1967, the first day back after a long, hot summer, so they'd finished at midday, just as they always did on the first day of school. Six-year-old Karen came home on the bus with her 15-year-old aunt, to her home in Strickland Avenue where the outer edge of Hobart extended up the slopes of Mount Wellington.[2]

It had been a smoky morning. As on many summer days in Tasmania, fires were burning across the state. Around Hobart alone, 110 separate fires burnt. Most of them had been burning for some time. Most of them had been deliberately lit. Only 22 of them were accidental. Some had been started to clear pasture or promote fresh growth. Some, flaring up under the dry, hot conditions, had smouldered in woodpiles lit weeks earlier. Seventy of the fires remained out of control on Tuesday morning.

The weather was not particularly bad that morning. It was hot but the breeze was mild. The fire danger index was high but not extreme.

The situation changed rapidly around midday, just as Karen arrived home from school. The wind picked up rapidly, gusting to 122 kilometres per hour in the afternoon. The temperature rose to 38°C. The fire danger index was now extreme.

The fires that were burning around Hobart flared in the hot, windy conditions, leaping and spreading through forests and open fields, roaring through townships, down gullies and around mountains. Where they met, they joined into huge fronts. With so many different fronts it was impossible to know where the fires were coming from or in which direction they would attack. One fire, which had started in Lime Kiln Gully in Glenorchy, was not regarded as hazardous earlier on but soon spread into surrounding bushland. It burnt across the Huon Highway, across Waterworks Road, and into Strickland Avenue, where Karen Schofield lived. At the same time, a fire had also been burning in Browns River for most of the day. When a third fire came over the top and around the sides of Mount Wellington, it pulled the other two fires together and combined to bear down on the suburb of Ferntree.

When the police told residents in Strickland Avenue to get out, Karen was bundled into a car by friends and neighbours, with their pets and possessions. They wanted to go up the mountain but were told the whole area was on fire; they'd have to go down towards the city. They didn't get far before they were confronted with flames. The fire had beaten them to it, there was no way out.

Karen Schofield's parents were in town. On hearing of the fires, they rushed up the hill to find their daughter. Many others also abandoned their desks to join the fire front, untrained and unprotected, but answering the call to arms just as Tasmanians have for generations. Karen's parents found their street blocked by a barricade. No-one was allowed in. They tried the other end. It too was closed. In the end, they just crashed through, determined to get to their daughter no matter what.

Strickland Avenue winds narrowly along the base of Mount Wellington. The Schofields didn't get far before their path was blocked again—this time by a fallen tree. Flames licked through the open car doors as they reversed back through the smoke. They found some men fighting the flames and asked who'd gotten out of the street. No-one, they were told.

'Wind your window up, lady, and get out of here,' Karen's mother was told.

The men were right. Karen hadn't gotten out. At the end of the street, the group she was with had been forced to abandon their car as flames approached and all 10 people crawled into a ditch, huddling together as the fire burnt around them.

While the devastation was widespread across Hobart, most of the deaths and house losses occurred in these suburbs. It was one particular fire out of the 110 that did the most damage, rather than the combined effects of all of them. The fire burnt to within 2 kilometres of central Hobart. More than 200 000 hectares of land were burnt, from the northern towns of Bothwell and Hamilton all the way to the southern reaches of the Tasmanian channel country. Agricultural losses were high. The air reeked with a characteristic sickly smell. Many small towns were completely destroyed. Fishing boats rescued terrified coastal residents as the fires swept down the steep hills. The boats floated just offshore and their occupants watched helplessly as their homes were engulfed in flames. The seaside town of Snug had just 40 houses left from 120.

Lines of communication failed. Power lines came down, telephone lines were cut, roads were blocked and the airport closed. Many radio stations were down and those remaining relayed frantic messages from survivors to missing family and friends. It wasn't until late in the evening that Karen's parents finally had word that their daughter was safe—salvation in the midst of tragedy.[3]

The Hobart fires shocked the nation. Tasmania was well accustomed to fires. From the earliest fires of 1897–98 and 1912, fires burnt regularly through the dense forests and open farmlands. Fires were normal, fires were expected, fires were part of rural life. But no-one had imagined that fires could stretch their reach into the heart of the city itself.

Hobart was the first Australian city to be burnt. It would not be the last.

SYDNEY 1994

Sydney is a city best seen from the sea. As you sail towards the heads of Port Jackson, you would hardly know that Australia's largest city lay just beyond these sandstone bluffs, with their smattering of native vegetation looking much as it must have done two centuries ago when James Cook sailed blithely past, unaware of the hidden treasures within. Cook had discovered the most southern of Sydney's inland waterways, Botany Bay in the south. It was the First Fleet that settled further north in Port Jackson but it did not take the early colonists long to spread north, discovering the third, and most extensive, of Sydney's major waterways—Broken Bay.

As visitors travel past the long, white beaches of Sydney's northern shores, around the lion's head of Barrenjoey into Broken Bay, they are faced with a wealth of choice. Turn north to tour the quiet, amiable waterways of Gosford and Woy Woy. Straight ahead leads into the agricultural heart of Sydney's hinterland, up the Hawkesbury River with its oyster farms and smattering of country towns. To the south, the bustling waterways of Pittwater take you directly to the midst of Sydney's busy northern suburbs. On the weekend there seem to be almost as many boats on the water here as there are cars on the road.

Sydney's wealthy suburbs hide their glamour discreetly beneath a curtain of green. Unlike some other states, which historically required land to be cleared of native vegetation before it could be developed, Sydney has a history of protecting its natural assets. In the northern suburbs, houses have had to be built around the trees and as much natural vegetation preserved as possible. Despite the massive density of housing, the hilly slopes of Sydney's northern suburbs are surprisingly green.

Not all of these green hills hide houses. Between the mouth of the agricultural Hawkesbury and suburban Pittwater lies another waterway, Cowan Creek. Almost completely encapsulated by the Ku-ring-gai Chase National Park, this waterway presents an image of what Sydney must once have looked like—a mass of calm, dark waters unfurling from a succession of densely vegetated sandstone outcrops. In the early light of a mid-winter morning, a blanket of mist rolls across the water's surface, stretching into the countless tiny bays, each carved from the rocks by rivulets of water tumbling down to the sea from their slopes. It is a spectacularly beautiful landscape. On a quiet day, it is an eerily ominous landscape, rich with an unspoken past.

The national park spreads back south, dark fingers stretching between the web of the outer suburbs. Sydney is surrounded by parks like this, from the vast expanses of the Wollemi, Blue Mountains and Kanangra Boyd national parks to the west to the smaller pockets of regional reserves. These parks encircle and contain the city but, as the city grows, many parks have themselves become surrounded by urban jungle, like the Royal National Park in the south or the Sydney Harbour National Park itself.

These parks are what makes Sydney so beautiful, what keeps it liveable and saves it from becoming the urban wasteland of so many of the world's large cities. It is possible to live in the very centre of Sydney, just minutes from the world of bankers and corporate intrigue, and yet be surrounded by bush and wake to the sounds of whipbirds cracking across the gully. It is at this interface, where people meet forest, that our cities are at the greatest risk from fire.

In 1983, a fire swept through Ku-ring-gai National Park, threatening the houses nestled between the water and the trees along the edge of Pittwater. I lived in Pittwater then and remember the hot winds of that day, the sun turning orange and the peculiar, dirty grey cast of the sky. Burning embers flew across the water, driven by fierce winds. We spent the day wetting down the decks of wooden boats, ensuring they didn't catch fire. This was only a small fire, largely contained within the confines of the park—an inconvenience rather than a direct threat.

Ten years later and parklands again caught fire. More than 800 separate fires raged across areas of the north coast and Hunter regions, the south coast and Blue Mountains, Baulkham Hills and Sutherland. Most of the Royal National Park was completely burnt, along with large stretches of the Blue Mountains and Warringah–Pittwater parks. Fires seemed to ring the entire city.

My father still lived in Pittwater then. He recalled the approach of the fire over the crest of Towlers Bay, a tiny collection of houses clustered on the water's edge, accessible only by ferry. The residents knew the fire was coming and knew they had nowhere to run. Some were prepared, ready to fight the fire with pumps and hoses. Others were not, clustered on the water's edge frantically clinging to overexcited children and overwrought pets. My dad brought his boat into the tiny jetties, as did many others, piling their decks with precious possessions, pets, goats, chickens and children, taking them out to the relative safety of the bay while others remained on shore preparing to defend their homes.

Fires usually burn up hills quickly and come down hills slowly. The people of Towlers Bay, at the foot of the hill, should have been defending from a slow, creeping fire. But topography can do strange things to fires. As the fire crested the ridge, it was as if it suddenly sucked in a deep breath from the fresh oxygen in the bay, changing the very direction of the wind. Roaring to a crescendo at the top of the hill, the fire swept down the slope with a renewed vehemence. Elizabeth Franklin, a resident in nearby Elvina Bay, recalled, 'I remember as we reached the wharf, hearing a wild roar, and turning round I saw the gum trees just keeling over sideways, before a seventy mile gale'.[4] Gum trees exploded into flames like great firecrackers, showering sparks all around them.

No lives were lost in Towlers Bay that day. The fire brigade vessel sucked an endless supply of saltwater from the bay, over 1000 litres a minute, dousing flames up to 40 metres away as they reached their ultimate termination at the water's edge. Furniture, goats, chickens and children were all safely unloaded and returned into the blackened landscape of their homes, shaken but unharmed. Ten homes were lost nearby in Lovett Bay and another two in Elvina Bay. Across the state, 800 000 hectares of land had been burnt, more than 200 homes destroyed and four lives lost. Images of Sydney ringed by fire were beamed across the world.

The risks of living near the bush, or even in the bush, are made all too obvious in these rare disasters. In high-risk areas, minor fires may occur every few decades, major ones every forty, fifty, even 100 years. For this one day in 10 000, the bush is a terrifying, life-threatening place. But for those other 9999 days, it is paradise on Earth, a garden of Eden. The risk of experiencing a fire is actually quite low but, the longer you live in a fire-risk location, the greater the probability that you will, eventually, be burnt out.

But humans are never good at calculating risk. There is a church on the Hawkesbury River that was flooded almost a century ago. According to locals, the organ from the church eventually washed ashore on one of Sydney's outer beaches. Such floods occur regularly, with years of major floods clustering every 40 to 50 years. Yet on the river's edge, well below this church, are two or three blocks of new homes, all built in the intervening years, all clearly due to be flooded, if not entirely destroyed, in some future flood. With clear evidence of past floods all around them, people still build here.

CANBERRA 2003

It is positively eerie to read the newspaper accounts of Canberra's 1939 fires. They seem so familiar. They could have been written yesterday. More specifically, they read like a prophecy of the fires that were to sweep through Canberra in 2003, 64 years later.

> The Minister for the Interior (Mr. McEwen) announced last night that he proposed to appoint a committee to study the experience of the week-end fires with a view to setting up a more effective organisation for combating bush fires.
>
> Mr. McEwen said that he had been impressed with the need for more effective safeguards of the valuable pine forests of the Territory and in particular of Mount Stromlo. He could not contemplate continuance of a situation in which Mount Stromlo was exposed to the fire risk revealed on Saturday. Canberra itself required more effective protection in the Black Mountain area, where there was a probable fire weakness.
>
> Mr. McEwen said that he proposed to take all steps necessary to assure the security of the capital, the safety of lives of residents in outlying parts, and protection of valuable Commonwealth assets in the Territory.[5]

Sixty-four years ago, Canberra was a city barely out of childhood, created from rural countryside just 12 years earlier. For all its planning and preparation, surely such a young city of just 10 000 people could be forgiven for being a little unprepared for fires of the magnitude that had taken the much older, more experienced cities of Melbourne and Sydney equally unawares?

For all of the minister's inquiries, 1939 was not the only year the Mount Stromlo Observatory was to be threatened by fire. It was not the only year the surrounding pine plantations were to pose a risk to life and property. In 1951, over 400 hectares of pines were burnt around Mount Stromlo, destroying several of the observatory's buildings and killing two people. On Christmas Day 2001, fires threatened Government House and the Mint.[6] With this history, how could the fires of 2003 come as such a shock to Canberra residents and Australians alike?

On 8 January 2003, dry lightning strikes caused over a hundred small fires to start in the national parks surrounding Canberra, particularly Mount Kosciuszko, Nagambi and Brindabella national parks. Such lightning fires are common in mountain forests and rarely pose a great risk to nearby urban centres. Very often they are left to burn themselves out, with firefighters concentrating on containment—limiting their spread out of the parks or into more risky areas.

Over the next 10 days, circumstances changed. The fires in the mountains continued to grow. The weather grew hotter, drier and windier. On 17 January, the strong winds pushed the fires well into the ACT itself. On the next day, a Saturday, they spread into the capital, with catastrophic results.

As early morning shoppers headed off to markets and families bundled kids into cars for the weekend sporting fixtures, burning leaves began to fall in the suburban gardens of Weston Creek, Kambah and Tuggeranong. The hot wind put some residents on edge and they spent the day clearing flammable material from around their houses. Others retreated to the cool air-conditioned comfort of their lounge rooms, drew the blinds and shut out the world.

Through the day the winds drove the fire front closer, stretching along a 35-kilometre front, reaching the edge of the suburbs by mid-afternoon. By late afternoon, the fire reached the pine plantations surrounding Mount Stromlo. Saturated with flammable turpentine and encased in fine needles of foliage, the pines went up like roman candles, creating a firestorm of unprecedented ferocity. Cyclonic winds effortlessly snapped 40-metre pines in half. Fifty thousand kilowatts of heat radiated from the pine plantations, which stretched from Mount Stromlo down to the houses on Warrangamba Avenue in the suburb of Duffy.

By the evening news, chaotic scenes of Canberra suburbs on fire were beamed around the world. In the prematurely darkened streets, people ran from houses to cars as fire trucks patrolled the burning streets and gale-force winds hurled burning embers through the air.

The fire damaged 70 per cent of the ACT's pasture, bush and pine forests. Within the space of just 10 hours, four people had died, and 500 homes and the historic Mount Stromlo Observatory were destroyed.

FROM CALIFORNIA TO ATHENS

Australia may be unique in the scale and frequency of its bushfires but its cities are not alone in facing the challenge of the burning bush. California has ecologically similar fire-prone vegetation. Like Australians, Americans are increasingly choosing a lifestyle at the interface between urban life and what they call the 'wildlands'.

Just as southern Australia moves out of its hot, dry summer, California heats up. Firefighters criss-cross the Pacific, the Americans coming to assist the Australians in January and February, while Australian contingents return the favour in August and September. We watch each other's fires from the comfort of our respective winters, the images from across the globe an unsettling reminder of what may face us next season.

Southern California is exposed to the Santa Ana winds, strong north-easterly winds created by high-pressure pockets of air trapped in the Great Basin between the Sierra Nevada and Rocky Mountains. In autumn, these winds bring hot, dry air sweeping down across the coastal fringe, raising temperatures and drying vegetation. Known also as the devil winds, or winds of *Satanás,* they have been associated with the worst fire disasters in southern Californian history.

The Santiago Canyon fire was an early lesson in natural hazards for the farmers and settlers of Orange, San Bernardino and San Diego counties. In 1889, after months of drought, strong Santa Ana winds fuelled a fierce fire that burnt across 100 000 hectares of chaparral and scrublands. No human lives were lost, but countless heads of stock were destroyed, along with crops, fencing and houses.

The Laguna fire of 1970, started by fallen power lines in strong winds, burnt across large areas of the state, killing eight people. The 2003 firestorm season comprised 15 major fires, including the Cedar Fire in which 15 people died and the Old Fire in which six people died. Five firefighters died in the Esperanza fire in 2006. The arsonist found guilty of starting the blaze was sentenced to death in 2009.

Despite dramatic improvements in firefighting capacity, the fires seem to get more tragic every year. In 2007, the Witch Creek Fire was part of a season of fires that claimed 14 lives and burnt over 200 000 hectares, the largest to affect the state. An arsonist suspected of starting one of the blazes was shot dead by authorities as he fled the scene.[7]

Lives are not being lost because of lack of resources or effort, or because the fires are bigger or worse than they were in the past. Lives are lost because there are more people living in the wildlands than ever before. The cities have overflowed into the country, bringing a population accustomed to the relative safety of urban life into a perilous environment for which they are ill prepared.

While California, like Australia, has a long history of bushfires, for other countries it seems to be a more recent hazard. The increasingly urban population in the countryside around Athens in Greece is seen as one reason for the increase in fatalities from fires, which culminated in the deaths of 70 people in 2007. Fires have always been a natural part of the Mediterranean climate, shaping the unique drought-adapted vegetation of the region. But the fires seem to have become more frequent since the 1980s and their consequences have certainly become more dramatic. In 2009, Athens was again ringed by fire, lighting up the hills behind the Acropolis in a spectacular, if ominous, display.

9 · THE BIG ONES

Every community has its own fire history, its own stories of survival, disaster and heroism. It is only when we delve into Australia's fire history that the depth of the detail is revealed; fires and seasons shatter and fragment into myriad smaller, local events, each reflecting the details of the larger story in finer resolution.

Emergency Management Australia maintains a disaster database of fires from throughout Australia's history. It has over 200 listed.[1] This is far from being a total record of Australian fire catastrophes. Small fires are not recorded, for all they may have destroyed the livelihood of a local community, burnt houses or killed livestock. Large events are clustered on a single line—fires that occur across states, across regions—and designated by a single epitaph, as if there was a single point of impact like a fully laden plane crashing to a fiery death, or a bomb blast. But fire impacts are inherently small, inherently local. Tragedy at one end of the street meets survival at the other. An inferno in one place adjoins unburnt bush in another.

Yet there are some events that exist beyond their local impact. We remember these events. We know what we were doing when we heard the news. We remember the weather, the reports, the footage. We have collected stories from friends and acquaintances, even when we weren't there. These events become part of our shared experience, part of being Australian. These fires fundamentally shift the way we think, the way we see our landscape.

Ash Wednesday 1983, the first day of Lent, was just such a fire.

ASH WEDNESDAY 1983

Wednesday, 16 February 1983, was all but over before the first hint of tragedy occurred. The nine o'clock weather bureau outlook that morning had forecast a maximum of 34°C for Melbourne—a fairly typical, hot, summer day for a southern Australian city. People sweltered on their way into work, anxious to escape into the air-conditioned comfort of offices and shopping centres. Children just back at school after the holidays ran around noisily under floppy sunhats, heeding the national 'Slip, slop, slap' message. Just another scorching day in an average Melbourne summer.

By midday, the predicted maximum had been increased to 39°C. By the afternoon, temperatures had soared to the hottest in half a century—43°C. A week earlier, a great cloud of desert dust, 500 kilometres wide and 100 kilometres deep, had swept across the city, depositing tonnes of topsoil from the drought-stricken inland along the coast, in the city and in the sea, turning the sky a dull, oppressive orange. Fierce northerly winds drove the stinging sands into the exposed flesh of discomfited city-dwellers. Humidity dropped to a dehydrated 6 per cent. The streets of Melbourne were a blast furnace. Out in the countryside, the piercing dry winds seemed to be actively seeking any unwary spark among the tinder-dry vegetation.

In small towns on the outskirts of Melbourne, old-timers shook their heads in the ominous heat. Long-time residents had seen these conditions before. The 1982–83 fire season had already been long and hard. Older residents of the north-eastern towns, like Warrandyte, Yarra Glen and Noojee, where the forested mountains stretch down to meet the outskirts of Melbourne, remembered the 1939 fires all too well. Even more people remembered the smaller 1962 fire that had swept from Christmas Hills into the Kinglake National Park and down to Kangaroo Ground. Many of the same men and women who saw their own homes lost in that fire now stood on verandahs listening on the hot wind for the wail of the dreaded siren.

The fire ban season was declared unusually early, on 24 November 1982, with serious fires already underway. By the end of January, consistent high temperatures had helped produce more than 30 fires across the state. The absence of high winds minimised the spread of those fires but not the damage—two forest workers died fighting a blaze fanned by gully winds in the rugged hills near Bacchus Marsh. In early February, when the massive dust storm swept the state, visibility was reduced to a few metres and rendered fire-spotting planes and towers useless. The state's early warning system was effectively out of action. Precious minutes and seconds were lost in responding to fires, and high winds turned small fires into towering blazes. One encircled Mount Macedon outside Melbourne; another burnt out over 100 000 hectares of forest near Cann River in Gippsland.[2]

The deadly combination of wind, heat and desiccation returned on Wednesday, 16 February 1983. A northerly wind gusted up to 60 kilometres per hour. A heatwave spread across South Australia and Victoria: 45°C at Coober Pedy, 43°C at Warrnambool, 39°C at Lakes Entrance. The entire south-eastern corner of the continent had been deprived of rain for weeks, sometimes months. The last decent rains anywhere had fallen 10 months earlier. Since then both Victoria and South Australia were classified as having experienced 'a serious rainfall deficiency'. Around the outskirts of Melbourne, rainfall for the past 10 months was the lowest on record. The region was a tinder box, just waiting for a spark.

Radio reports came in that large parts of South Australia were ablaze. Fires had started across the state, in the Clare Valley north of Adelaide through to the forests of the southeast corner. By midday, they were all out of control. Fires devastated the small historic townships of the Adelaide Hills, barely recovered from the last bushfire disaster just two seasons earlier. The fires respected no social boundaries. The mansion homes of the well-to-do burnt as quickly as the caravans and shacks of the less well-heeled. Priceless art collections and historic artefacts literally disappeared in a puff of smoke. Throughout the afternoon, Adelaide reporter Murray Nicol reported live on-scene as his own house in the Adelaide Hills was engulfed and destroyed.

> We are in major strife. We are crouching down behind a farmhouse. There are children. They're spraying us with water. The sky is red and then white. It's going crazy. The fire's jumped 100 feet high, 150 feet high, straight over the top of Greenhill Road. There are something like 120 houses at risk up here. I don't know how many have gone. We are in deep trouble. We don't know... we can't see any houses. There are a dozen people here with me. We can hardly breathe. The air is white. The air is white with heat and smoke and it's red and there are women crying and there are children here and we are in trouble. The air's full of cinders and smoke and I can hardly breathe. There are people crying and this is just too much. And I can't... I really can't believe it's happening. The wind here is gusting 50, 60 mile an hour and none of us can see. A declaration of

> disaster is in force throughout South Australia. At the moment I'm watching my house burn down. I'm sitting out on the road in front of my own house where I've lived for 13 or 14 years. And it's going down in front of me. The roofs are falling in. It's in flames and there's nothing I can do about it. Absolutely nothing. There's been a fire unit here from Summertown pumping water till we ran out and the flames are in the roof and, oh, goddammit, it's just beyond belief. My own house! And everything around it is black. There are fires burning all around me, all around me. And the front section of my house is blazing. The roof has fallen in. My water tanks are useless. There is absolutely nothing I can do about it.[3]

His distraught broadcast was heard across the world.

But despite the flammable conditions, by mid-afternoon no significant fires had been called in for Victoria. Residents in high-risk areas held their breaths. All around the state, volunteer firefighters congregated nervously near fire stations, knowing that a few seconds' delay could see the smallest fire whipped up into an uncontrollable fury by the relentless winds. The phones stayed silent. Would they make it through the day into the cooler, less dangerous night conditions without a fire?

It wasn't until three in the afternoon that the first banshee wails of the fire sirens sounded across Victoria. Fires in the southwest, east and northern districts started in quick succession. A small but fierce fire had been spotted moving rapidly across Birds Paddock, a heavily vegetated nature reserve in the Dandenongs, just near Belgrave outside Melbourne. In the dry conditions, such fires seemed to light spontaneously. Their all-too-frequent distribution around the edges of cities and towns suggested a far more human origin. The fire in Birds Paddock was away from roads, power lines or machinery that might be the source of an accidental spark. No evidence of lightning strike was ever found, nor glass that might have intensified the sun's rays. The fire must have been started by someone, but how or why, accidental or deliberate, is a mystery.

Belgrave South Fire Brigade took the fire call at 3.23 p.m. and within moments trucks were speeding towards Birds Paddock.[4] By the time they arrived, just a few minutes later, the fire was already out of control. The fire had started 'crowning'—snaking up into the canopy and setting alight the very treetops. The wind blew embers and sparks up to 10 kilometres ahead to start fresh outbreaks, enabling the fire to easily leap roads and highways that might have acted as a break for a more manageable ground fire.

Assistance from nearby Toomuc Valley Fire Brigade was sought. Tankers from Upper Beaconsfield Brigade turned out, sirens wailing and lights flashing, moments later. Within two hours the area was teeming with tankers and firefighters from within the local Pakenham group of brigades. Before long they were joined by tankers from the neighbouring brigade groups.

Fighting a fire that travels across the ground and through the canopy at a brisk walking pace is practically impossible. Unable to approach the main fire front, firefighters and their resources were spread across the region attempting to protect homes and property, defending people sheltering in schools, kindergartens and hotels, and attempting to control new spot fires as they emerged. Roads were congested with emergency and utility services, evacuees, friends and relatives. Some people raced into the area to protect their homes. Some left to protect their families. Most did nothing, not realising the peril they were in.

One of the first casualties of the fire was the electricity

supply. Burnt power poles required the electricity authority to cut power to the entire area, including to the water pumps that filled the storage tanks providing water to local households and fire hydrants. Unbeknownst to the desperate householders and firefighters drawing on the precious remaining water, time was running out.

Strike teams of firefighters and tankers from neighbouring brigades and regions were hastily assembled in stations further afield and dispatched to assist the affected area. Soon brigades from regions all across the north and east of Melbourne were asked to provide a crew and tanker to help in the fight, while still leaving enough volunteers to protect their own district should new fires strike. Small towns with populations of just a few hundred sent tankers and crews, praying that they would not be needed closer to home in the meantime. Soon tankers and crews from towns 50 kilometres away arrived, their response to the emergency slowed only by the drive to reach the fire.

Firefighters often look forward to the evenings, when the cooler and generally moister air calms the flames and makes control work a little easier. But not this evening. The entire afternoon had been shrouded in a dark, oppressive haze, lit only by a fiery red sun engulfed in smoke and outshone by the raging fires. The impending nightfall was barely detectable. Just before eight o'clock a second fire—deliberately lit by a young teenager—was reported at Cockatoo, just north of Upper Beaconsfield. Resources already stretched in Upper Beaconsfield had to be diverted to protect the town of Cockatoo. Radio communications attempting to coordinate the activities of hundreds of tankers now had to cope with two major fires. A total of 121 brigades were eventually deployed to fight the Upper Beaconsfield and Cockatoo fires, nearly all crewed by volunteers.[5]

No way forward, no way back

The Panton Hill crew was led by Bill Marsden, an experienced lieutenant in the brigade, supported by two other senior firefighters, Peter Singleton and Maurice Atkinson. Between them they had 60 years of firefighting experience. The two younger brigade members were no strangers to firefighting either. Nineteen-year-old Neville Jeffrey had followed his father into the brigade, while eighteen-year-old Stuart Duff was a third-generation member. But nothing could have prepared them for the inferno they were about to face.

Back in Panton Hill, friends listened anxiously to radio traffic from the fire ground. A former captain heard the Panton Hill crew asking for directions in the unfamiliar territory, before they followed a local crew from Narre Warren up a steep, heavily wooded, fire-access track in Walkers Gully, leading to St Georges Road. Houses in St Georges Road were at risk from the fire and the two tankers were on their way to defend them.

The Narre Warren tanker had been on the fire field since the outbreak in Birds Paddock. Relieving their exhausted colleagues, this second crew had taken charge of their brand-new International tanker, delivered to the brigade just days before, knowing they had the latest in firefighting equipment at their disposal. Their crew leader was brigade captain John Minett, a highly experienced firefighter, awarded Fireman of the Year a few days earlier. John had conducted exercises in the Upper Beaconsfield area before, along with some of his crew, but in the dark, blinding smoke, even this limited local knowledge seemed worthless. His radio operator, Dorothy Balcombe, one of the few women in the brigade to take an active role in firefighting, stayed glued to the radio—their lifeline to the outside world.

The two tankers headed up the narrow, unfamiliar track leading into the forest, not knowing what lay at the end, whether it offered safe clearings to shelter in, room to turn and retreat, or alternative escape routes. They disappeared into the smoke-filled forest just as the first gusts of a cool southerly change arrived.

Back at control headquarters, a relief crew from Panton Hill had arrived by bus to take over from their tired colleagues. But no contact had been made with the Panton Hill tanker or the Narre Warren tanker for some time. The last official contact was logged at headquarters from Dorothy Balcombe on the Narre Warren tanker just before nine o'clock.

'Narre Warren Tanker One in Upper Beaconsfield lane. Desperate situation.'[6]

At about the same time, someone thought they heard the Panton Hill tanker.

'Mayday, mayday, we can't go forward and we can't go back.'[7]

When dawn finally broke on that long, long night in hell, 21 people in Upper Beaconsfield alone were dead, almost a third of the final Ash Wednesday death toll across two states, and the distant townships of Upper Beaconsfield, Narre Warren and Panton Hill were united by grief and tragedy.

Seventeen firefighters died in the Ash Wednesday fires, including the crews of the Panton Hill and Narre Warren tankers: fourteen in Victoria and three in South Australia.

Strike teams had a clear focus on the task at hand. Although they were working in unfamiliar terrain, they had the benefit of knowing their own properties and loved ones were not at immediate risk. Many of the local firefighters had to face both the fires and the fear of losing their homes and leaving their families to fend for themselves. For some, the fear was well founded. Many residents were poorly prepared for the fires, favouring last-minute, often disastrous evacuations on smoke-choked roads blocked by stalled traffic, fallen trees, escaped animals and other hazards. Few had appropriate clothing and fire plans beyond using a garden hose. Many residents believed that 'someone' would evacuate them if the situation became dangerous. For too many, that 'someone' never arrived.

The risks to residents, inadequately prepared for the inferno that approached them, were just as great as the risks facing the firefighters on the fire ground—better prepared but in far more dangerous conditions.

The freshening winds from the south, which arrived around 8.30 p.m., seemed to offer welcome relief from the blazing heat of a day on the fire ground, but any respite was short lived. This change in weather brought no welcome rain or moist, still air to cool the fire's ardour. The night winds may have been cooler but they were also gale force, breathing fresh life into the fires. Over the course of the day, the fire had spread rapidly south-south-east along a narrow corridor 5 kilometres wide and 16 kilometres long. Fuelled by the southerly gale, the extended eastern flank of the fire suddenly became a new, bigger and even more deadly front, roaring northward into new territory, overtaking horrified firefighters and residents alike. Despite weather bureau warnings of the impending southerly change, news had filtered through to the fire ground barely half an hour beforehand. Few residents were even aware of the approaching change and those who were failed to appreciate its significance.[8] There was no time to redeploy firefighting resources; there was barely enough time to get crews and householders out of harm's way.

As the fire front approached, few would see the full scale of its ferocity. Eyes were clamped grimly shut against the stinging smoke and debris. Showers of burning embers rained down on bare, summer-clad flesh of ill-prepared residents. Debris from the fiery maelstrom was flung by the gale-force winds, smashing windows, walls and roofs, driving burning brands deep into homes. The acrid stench of remains—animal, vegetable, mineral—impregnated everything. The heat, the overwhelming, inescapable heat, was unbearable—the final overpowering sensation for far too many victims. The only light at the end of that dark, infernal night was the glow of the approaching flames.

For the survivors, it was the sound of the fire that truly terrified them: the vast, unstoppable, deafening roar of an unleashed beast, leaving nothing and no-one unscathed in its path. All that stood between those who barred the way and death was a thin trickle of water from the end of a hose.

At the peak of the inferno, the last of the water in the supply tanks ran dry. Ash Wednesday, with all its sombre religious connotations of mourning and grief, lived up to its name that year.

KINGLAKE 2009

I guess no-one really believed that a tragedy on the scale of Ash Wednesday would happen again. There had been so many improvements in our understanding of fires, so many improvements in how we fight fires, so many improvements in communications and technologies. Surely nothing could be that bad again?

Even those who predicted the worst didn't imagine it would turn out as badly as it did.

Some years are bad fire years. Some years, the fire calls start before winter is over. Some years, it seems that barely a weekend goes by without the pager bleeping madly and the frantic dash to the station. In some years, the wet spring cultivates a thick tangle of waist-high grass, which dries wheat gold in the sun.

But 2009 was not one of those years. Victoria had been more or less in drought for the last decade or so, and much of Victoria had experienced below average rainfall since 2005 in particular. Water storages for Melbourne hovered below 40 per cent, steadily receding with every summer. The spring rains had all but disappeared, the idea of lush, thick pastures receded into distant memory. The trees had shrunk, retracting their canopies as they produced fewer and fewer leaves with each dry season. From the air, what was once a uniform solid canopy of green turned mottled grey and brown, bare earth visible between the struggling trees.

What fuel there was, was dry. Firefighters dismissed the risk of serious grass fires for the season—there was no grass to burn. The season was neither unduly hot nor unduly windy. There had not been any more fires than normal. But it does not take long for a normal summer, even a mild summer, to turn into a nightmare summer. Just a few weeks with no rain, just a few days of heat, a bit of wind, and the conditions turn from average to devastatingly treacherous. These were the conditions on Saturday, 7 February. In late January, a heatwave struck the state, with temperatures around the state soaring to over 40°C for days on end. A large high-pressure system sat over the Tasman Sea, while an intense tropical low-pressure system off north-western Western Australia and the monsoon trough blew hot air from the tropics, drying out over the desert, south into Victoria.

Saturday was the worst day of all. The temperatures climbed to nearly 47°C in Melbourne—one of the hottest days on record. People were warned to keep cool and hydrated but hundreds of people collapsed and died in the heat. Fire warning messages blanketed the media. People in the country and on urban fringes were warned to be on high alert. People in the city were warned to stay in town.

The fire danger indices were extreme across the state. On a scale where over 50 is extreme, most locations were recording in the hundreds. The peak was at Ballarat, with a grass fire danger index of 330. A northerly wind of 50 to 60 kilometres an hour blasted its hot breath across most of the state.

Nerves were already fraught. A fire had been burning at Delburn, south of Moe in Gippsland, for the previous week. The fact that it seemed to have been deliberately lit did nothing to ease anxieties. Arson also seemed responsible for a second fire starting in early February at nearby Churchill, threatening the towns around Callignee. At the same time, a major fire had started in the Bunyip State Park and, despite the best efforts of the Department of Sustainability and Environment staff during the cooler days, it still posed a significant threat. If the fire got into the more rugged hills, it would prove impossible to control.

Brigades around the state stood by on alert, ready to leap on any fire before it had a chance to get away. It would only take minutes, in the wind and heat, for a spark to become out of control, for a wisp of smoke to become an ominous plume. By and large, the brigades were successful. Hundreds of fires were extinguished before they had a chance to grow. In the Western District, firefighters put in herculean efforts to bring grass fires under control but their success was shadowed by disaster elsewhere.

Just before noon, the Mount Hickey fire tower reported smoke a few kilometres east of Kilmore, directly to the north of Melbourne. Kilmore lies at the north-eastern tip of a band of mountain forests that extend around the northern and eastern perimeter of Melbourne. These forests rise from Kilmore up through the Kinglake Ranges, round the top of the Yarra Valley and around into the Dandenong Ranges. It was as if the fire started at the mouth of a funnel leading into some of Victoria's most flammable vegetation.

A fallen power line seemed to have sparked a grass fire on top of a hill. The local brigades of Broadford, Kilmore, Clonbanine and Wandong responded within minutes, but in the high winds the fire raced through the grass, pushing it rapidly towards the nearby forests. Tankers sped from surrounding towns towards the scene, knowing they only had minutes to bring the fire under control. It was to no avail. The fire was too big, too fast, to be contained and, despite the firefighters' frantic efforts, it roared away to the south-east, into the forests, under the power of increasing winds.

Some time after noon, more fires started on the Glenelg Highway near Coleraine and at Horsham in western Victoria. The onslaught had started. Everyone hoped against hope just to get through the day without too many losses. A cool change was due to arrive late in the afternoon. If they could just keep the fires down, put them out early or restrict their spread, tomorrow would be a better day. If the fires spread and expanded, the cool change would spell disaster, producing an extended front from the ever-lengthening sides of the fires and fuelling it with fierce southerly winds. They'd seen the damage cool changes could do on Ash Wednesday and in many other fires. No-one wanted to see that again.

The early signs were not good. By late afternoon, the residents of the Kinglake Ranges and foothills were overwhelmed by flames. Mayday calls came in from tankers trapped as the fire front burnt over them. They had answered a call to save residents and now struggled just to save themselves. Residents and firefighters sheltered together in their trucks under a fine mist of water and hoped that their training and equipment would save them. Calls to 000 overwhelmed the service, plotting the path of the fires as residents clung to their telephones in the desperate hope that someone could save them.

At three o'clock in the afternoon, the fire tower at Mount Despair saw smoke rising near the Murrindindi Mill. The smoke could soon be seen from another fire tower at Mount Gordon, just near Marysville. The men and women who stand guard in the fire towers throughout the summer are a remarkable breed. They know every lump and bump in the landscape. They can triangulate the origin of a puff of smoke rising from a blanket of forest to within 100 metres. They can see the dust raised from the car of an arsonist speeding away down a dirt track. Very often they can tell the difference between a car fire, a bonfire or the stale, drifting smoke from a distant control burn. And they can tell when a fire is the very devil on a rampage.

Andrew Willans, the tower operator at Mount Gordon, a veteran spotter of some 18 seasons, described what he saw coming towards him that day:

> That cloud mass from its infancy in that first half hour, it was huge. In that first half hour, from a tiny little trindle of smoke that I spotted and was able to give a bearing to, it grew in such incredible stature. I'd seen nothing like it. It terrified me. In the next half hour to when it breached the Black Range Road, I guess it would have doubled the amount of smoke, its height, its width. It was enormous. The following half hour until I left it was massive. It's beyond description. This thing was huge, absolutely huge. I can't explain it.

> And it was alive. This thing was just full of ember, ash, burning materials. This thing was absolutely alive.[9]

Fires spotted kilometres ahead of the main front, first one or two, then dozens. By half-past four, the tower operator evacuated, speeding back to defend his own home. Soon afterwards, the 9-metre-high tower was destroyed as the fire consumed all in its path.

At five o'clock a cool change swept through Melbourne, blowing away the day's record-breaking heat and bringing relief to the city. At around half-past six, the cool change blasted through the foothills and mountain forests, sweeping the fires through the town of Marysville within minutes. Early reports suggested that everyone was safe, that the town had congregated on the oval. Many people had. But a great many more had perished. Out of 400 houses, only 15 remained standing.

It took weeks for the fires to be finally extinguished. Even after the horror Saturday, the fires continued to threaten properties and townships. New fires started on Wilsons Promontory and French Island. Every hot day, every strong breeze, threatened to set the fires raging again. Everyone stayed on edge, traumatised by the disaster, the lives lost, the uncertainty of life.

Not until March did the rains come. Summer was over. Everyone breathed a little more easily and turned to the painful task of grieving, taking stock and thinking about the future.

WHAT NEXT?

There is little doubt that history, at least in relation to fires, repeats itself. Perhaps not on a scale of years or decades, but it certainly repeats over a scale of centuries. The conditions in Victoria on 7 February 2009 were worse than they had been on Ash Wednesday in 1983 and worse than they had been on Black Friday in 1939. But the descriptions are strikingly similar to those of the 1851 fires, just a few years before official weather records began.

There is no reason why vast or disastrous fires should not occur again. We hope they won't. We imagine that we are better prepared, better organised, better protected. We are not. There are more of us, at greater risk, now than ever before. There was no reason why the extreme conditions of 1851, 1939 and 1983 in south-eastern Australia should not recur. And recur they did in 2009.

It is long past the time to accept that despite our best efforts we cannot, in these rare extreme conditions, control nature. We cannot control fires in severe conditions or prevent them from happening. There is no reason why the huge inland fires of 1974–75 might not recur, nor any reason why a fire of the scale of the Peshtigo tragedy might not recur in the United States. Fires should not take any Australian by surprise.

What we can do is recognise the inevitability of large and catastrophic fires—and be better prepared to survive them.

Major fire areas in south eastern Australia

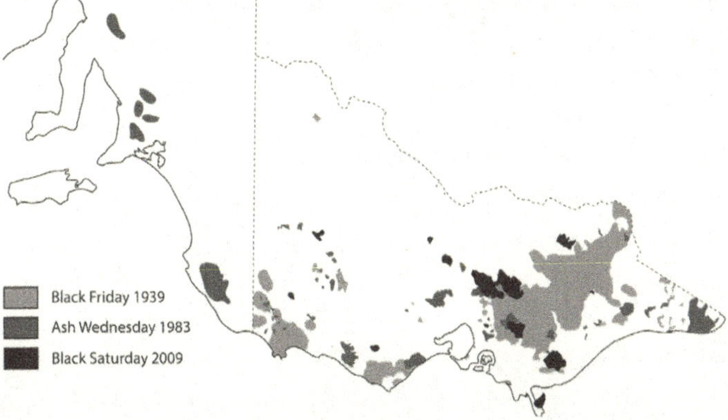

III · THE FIRE FACTORS

10 · IGNITION

> *A bushfire is not an orderly invader, but a guerrilla. It advances by rushes, by little venomous tongues of fire in the grass; it spreads by sparks burning leaves and bark. Its front is miles deep. It is here, it is there, like a swarm of venomous wasps. It shams dead and stabs you in the back. It encircles you so that there is no sure line of flight of its intended victims. It destroys the bridges in your rear. It bars the road with blazing trees.*
>
> HG Wells, Canberra, 1939[1]

THE NATURE OF THE BEAST

Australians have had a long and turbulent history of exposure to fire. Fire is a near-universal risk for rural communities in all but the wettest and driest areas. With our major cities clustered in the areas prone to some of the fiercest fires, our urban centres face a particular threat. There are few communities unaffected by fire, few people untouched, few areas without a significant fire history. Yet our understanding of this fearsome phenomenon has only really developed in recent years.

In essence fire is a chemical reaction, created in the ménage à trois of oxygen, fuel and heat (or a fire triangle, in the more prosaic language of firefighting manuals). Many materials can provide fuel for a fire but we can start with the one that concerns us most in bushfires—vegetation or plant material.

As the vegetation heats up, changes occur to the cellulose that makes up the plant. At about 150°C, the chemical bonds holding the cellulose together break apart, releasing unstable gases, otherwise known as smoke. Essentially, the electrons that orbit the molecules in the plant become excited by the heat and jump around, causing some of them to spin away from the molecules to which they are normally attached. As the temperature rises to around 260°C, these gases combine with oxygen in the air, glowing with heat and light as they do so, and we see them as flames.

Once a fire gets started, it no longer requires additional heat to maintain the chain reaction. The chemical reaction itself produces enough heat to be self-perpetuating—more heat creates more gases, which burn and produce more heat. If we can cool the fire down sufficiently that the heat it produces is no longer enough to sustain the process, then the fire can be extinguished. This is usually achieved with water.

Other techniques for extinguishing fires rely on removing other components of the fire triangle. Fire blankets can be used to smother small fires, depriving them of the oxygen they need. Foam, carbon dioxide and halon extinguishers also work by replacing the oxygen around the fire with other substances, suffocating the flames.

Oxygen is a vital element in the fire triangle. It was only when plants produced enough oxygen in the atmosphere that combustion become possible, and fire was born. More oxygen in the atmosphere, and combustion would occur at lower temperatures, burning up the planet. Like the ancient gods, oxygen gives us life but also carries the power of destruction.

In a big fire, there is very little we can do to control either oxygen or heat. The only thing we have any control over (and even that only slightly) is fuel. Backburns and fuel breaks are some of the primary means of bringing a fire under control.

When we deprive it of new areas of fuel to move into, the fire will eventually burn itself out.

But before we consider what sustains fires, and how to put them out, we should consider how they start in the first place. Who or what is it that provides that initial application of heat—that first ignition point that starts the whole chain reaction in motion?

GETTING STARTED

Before humans came to the continent, lightning was the only source of ignition for fires. And a very effective source it was too. Several hundred million volts course through the atmosphere with every strike, heating the air to 30 000°C and just looking for a target on which to vent their pent-up energy. Even in pouring rain, a lightning strike can set fire to an old hollow tree, leaving it smouldering and steaming in the wet. In the dry, lightning strikes are incendiaries, setting fire to hilltops and mountain ranges.

The pattern of lightning strikes in Australia closely mimics the occurrence of bushfires across the continent. The most frequent strikes are in the north, in late winter and early spring, coinciding with the dry thunderstorms responsible for starting so many of the cool, quick fires that burn across the northern savannahs.

As Australia's climate dried from the warm, wet world of the dinosaurs, fires became a regular part of the environment, shaping the vegetation and animals that evolved here. When humans arrived, they brought with them a new source of ignition. While nature probably remains the biggest pyromaniac overall, humans run a very close second.

In Victoria, a quarter of all bushfires are started by

lightning, the single largest source of ignition. About 20 per cent of bushfires are started deliberately—by illegal burn-offs, dumped and torched cars, children or arsonists. Escaped private burn-offs account for 18 per cent of fires, with a further 10 per cent starting from camp-fires and barbecues. The remaining fires have a variety of sources: cigarettes, matches, machinery and exhaust. The causes of some fires are never known. Despite their bad reputation, escaped prescribed burns start only 2 per cent of Victorian fires, at least according to the statistics provided by the department responsible for managing controlled burns.[2]

Not all fires are dangerous. Not all ignitions are equal, in terms of the area burnt or the damage done. If we look at the area burnt by fires, lightning is responsible for half of the area burnt by bushfires in Victoria every year. Lightning often strikes in remote mountain parks, where fires are both more difficult to control and less likely to do damage. There is less capacity to control these fires, and less need to do so. Fires started by lightning cause significant economic damage, particularly to forestry, but are responsible for relatively few lost houses, deaths and injuries compared with the areas they burn.

The next most significant source of fires is public utilities, a category that includes both trains and power fines. Despite the fact that these fires account for just 2 per cent of all ignitions, their impact is disproportionate. This single category of ignition accounts for 14 per cent of the total area burnt.[3] These points of ignition are likely to go unheeded or unnoticed for some time, allowing the fires to become established before efforts are made to extinguish them.

The vast majority of 'public utility' fires are actually caused by power lines. Unlike lightning strikes, power-line fires tend to occur closer to populated areas. House loss, fatalities and injuries tend to be much higher from these fires than from fires started by lightning strikes. Power fines have been implicated in many of our most devastating fires. On Ash Wednesday, the fires that caused so much devastation in the Adelaide Hills originated from power lines. Four of the eight major fires in Victoria that day were also thought to have been started by power lines.[4] Near Hamilton, in northern Victoria, a snapped power pole is alleged to have caused a fire that killed one firefighter. That power pole had been condemned two years earlier but never replaced.[5] On Black Saturday, a power line is alleged to have started the fire at Kilmore that killed more than 100 people and destroyed over 1000 homes.[6]

Power-line fires are relatively infrequent, yet they exact a disproportionately high toll in relation to house loss and fatalities. They are caused by a readily identifiable and potentially manageable source. This is a clear example of a high-impact fire cause that could be reduced through improved management. After he lost his house in Ash Wednesday, the chair of the South Australian Country Fire Service, Peter Schwerdtfeger, pointed out that the cost of the damage to the Mount Lofty area would have paid for putting most of its power lines underground.[7] In the Victorian Bushfire Royal Commission, a power authority argued that the cost of putting country power lines underground would have been $20 billion in 1990.[8]

Not all power lines are equally hazardous. Putting them underground is not the only way of reducing the risk. Power-line fires should be a manageable and declining source of bushfires. Standards and practices should be constantly improving and the risk they pose to the community should be reducing. Instead, many people seem to feel that the standards of power-line maintenance and management are declining. Self-regulation and privatisation in the electricity sector may, arguably, make good economic sense but it seems very unlikely to produce the improvements in safety we need.

Power to burn

For centuries, humans have set fire to themselves, their homes and their cities through the use of domestic fire. Lanterns, candles, fireplaces and woodstoves have long been sources of both comfort and occasional catastrophe. The development of gas lighting did little to improve this risk factor, adding instead a highly explosive element to the mix.

The development of electricity in the late 19th century was widely heralded as the safe alternative, reducing the risk of fire and explosion to negligible levels. Most cities and larger towns in Australia were electrified by 1905 with most small towns being connected by the 1960s.

Despite obvious safety improvements with electricity, electrical faults and appliances in homes are the most common cause of ignition, accounting for around a third of all house fires. The culprits, in the vast majority of cases, are appliances like faulty electric blankets shaken out from summer storage or a bar radiator too close to clothes drying on a rack.

In the silence and solitude of a peaceful bushland setting, electricity seems to be an unlikely source of ignition. But the distribution of electricity across the continent comes at a cost. Produced in vast industrial centres, close to their source of raw material, electricity is transported live through isolated, low-resistance wires. Where the electricity encounters resistance it gives off some of its energy in the form of heat or light:

in a radiator, or a light bulb, or occasionally through an errant possum or tree.

Fires can start in a variety of ways. Power poles can be snapped by the wind or by a traffic accident, bringing the lines in contact. Trees may grow too close to lines, particularly on private property, where they may not be cut back. In hot weather, power lines expand, sagging in the heat. They usually retract and tighten on cooler nights but, with successive days of heat, the air may not cool enough for the lines to contract, causing them to sag still further. In high winds, the loosened lines swing into contact, sending showers of sparks into the dry surroundings.

ARSON

Escaped burn-offs, camp-fires and other deliberately lit fires burn much smaller areas than their frequency would suggest. These fires are lit directly by someone, and that someone is, perhaps, also more likely to try to extinguish the fire or alert authorities, once they become aware of the problem. Escaped burn-offs, however, are responsible for high levels of injury and deaths, perhaps because of their proximity to people.

Deliberately lit fires are a strange category, which includes bushfires that are started for other reasons (burning a stolen car, for example, or burning down a shed in an attempted insurance scam). They may also include illegal burn-offs—maybe someone frustrated by the lack of backburning on a neighbouring reserve. This category also includes fires started by children playing with matches—a significant source of house fires but a less-frequent cause of bushfires.

The source we all tend to think of in this group, however, is the arsonist. When the Institute of Criminology recently reported that half of all bushfires with known causes are deliberately lit,[9] some leapt to the assumption that most bushfires are caused by arson. Arson grabs the headlines and sparks outrage and fury. Maliciously lit fires tend to occur close to population centres and so tend to be noticed more quickly. They are also more likely to be lit on high-risk days, when fires have a greater capacity to get away and do more damage.

Traditionally, arsonists have been defined as having educational or social problems but it has to be noted that these are the arsonists who get caught. A great many more don't get caught; this group may be quite different. But, in general, arsonists tend to be young men in search of excitement.

Fires are violent, powerful and aggressive. On a hot day, they explode into uncontrolled fury. Fires cause immense damage and hardship for a large number of people and yet, they are incredibly easy to start. Armed just with a cigarette lighter or a box of matches, a powerless, socially alienated person can unleash an enormous, destructive force across the landscape, at little risk to themselves and without having to face their victims. Fires are a weapon even the most disenfranchised can wield with great effect. Perhaps it's just fortunate that so few people choose to do so.[10]

Another popular misconception is that most arsonists are firefighters. With over 200 000 volunteer firefighters across Australia, it is not surprising that some arsonists have infiltrated the service. When they are caught, or even when they are suspected, the story nearly always hits the front page. But the numbers involved are small indeed. In a targeted investigation of suspicions or maliciously lit fires in New South Wales, 11 volunteer firefighters were found to be involved, out of a total of 50 people charged in relation to 16 000 fires. From a force of 70 000, that is a very small proportion of rotten apples indeed.[11]

Arson does seem to be on the increase but that may simply be due to our increasing population. While arsonists cause a great deal of damage and tie up a significant level of resources each summer, the greatest damage they do is in dividing our communities and breeding suspicion, fear and anger. Dealing with the underlying causes of arson seems to involve building stronger communities, keeping our youth occupied and involved, and by engaging those on the fringes so that they have a sense of belonging—something to lose.

HUMAN AND NATURE?

Statistics on the causes of bushfires are inherently variable across states and across countries. In some areas of America, lightning seems to cause the vast majority of fires. In Australia around half of all bushfires seem to be natural in origin, although some places, particularly near cities, may have a much higher proportion of fires started by people. How does this fit with the historical evidence that humans do not seem to have increased the overall incidence of fire in Australia, either through Aboriginal occupation or European colonisation?

We may well be good at starting fires but we are also good at putting them out. Suppression activities are probably quite effective at extinguishing small fires more rapidly than they would naturally be put out. Our efforts may not have much impact in terms of changing overall levels of burning, but we can change the time and place that burning occurs. What seems to have happened is that we have reduced the level of burning in some natural ecosystems while increasing it in others. More importantly, we have probably increased the level of burning around our homes, in the urban fringes of our cities where we are at our most vulnerable.

Fires only burn where there is fuel. No amount of lightning will fire a desert if there is no vegetation to burn. Humans, or lightning, can provide the spark, but how much a fire burns, how hot it burns and how far it burns depend not on the source of ignition but on fuel. How much vegetation there is depends largely on how much rain there has been for it to grow and then how little rain there has been for it to dry. Climate, not humans, determines how flammable our land is, because climate, it would seem, determines our vegetation.

As much as climate dictates the nature of our vegetation, humans certainly do determine the extent and nature of that vegetation. Two centuries of European settlement have brought about radical changes to the Australian landscape. Our forests have shrunk dramatically; our native grasslands have been replaced by pastures and crops. But the remnants of native vegetation, and some of the introduced varieties, remain just as fire prone as ever. Is it possible to protect what is left, to live in close proximity to it, while at the same time protecting ourselves from the inherent dangers of the bush?

11 · FUEL

BURNING THE BUSH

When we choose to live in a forested area, we are taking on a level of risk that is not faced by our more suburban friends. There is no way of removing this risk, short of moving to the centre of a concrete jungle. The methods of reducing risk are many and varied but they will never guarantee absolute protection from bushfires.

Bushfires are called bushfires (rather than wildfires) because that is what they are—they are primarily fires of the bush. We can be pedantic and discriminate between grass fires and forest fires. There are some types of non-native vegetation that carry a particular fire risk, like pine plantations, but broadly speaking, bushfires are a feature of the dry Australian bush—whether that be the tall forests of south-eastern and south-western Australia, the low mallee shrublands of inland Australia or open grasslands.

The closer we live to the bush, the more at risk we are from bushfires. Pretty much the only exceptions to this in Australia are the rainforests of north Queensland, wet western forests of Tasmania and desert country. If you live near a large area of dry bush or vegetation, even long grass or pasture, you are at risk. The bigger the area of bush, the higher the risk. The thicker and taller the vegetation, the bigger the risk. If there has been a fire anywhere near your area, in vegetation similar to where you live, at any time in the last 200 years, you are at risk. If you live at the top of a steep, north-facing hill, overlooking great swathes of tall forest, your risk factor goes through the roof.

The closer you are to the bush, the greater the risk of your house burning down. Studies of three major bushfires in the eastern states (Canberra 2003, Ash Wednesday 1983 and Sydney 1994) found that all of the houses that were lost were within 700 metres of bushland.[1] My house, which nestles within 8 hectares of low stringybark forest and adjacent to a forested reserve, faces a 60 per cent chance of burning down if a bushfire should strike my area. These figures are particular to the risk posed by forest fires, rather than grass fires, but given the proximity of many of Australia's major cities to bushland, this risk still affects a considerable number of people.

According to these studies, 4 per cent of residents in our capital cities (excluding Darwin) live within 50 metres of bushland—334 000 households in total are at the highest risk from bushfires. Twenty per cent live within 700 metres of bushland—that is, more than 1.5 million Australian households live in areas at risk from bushfires in our capital cities alone. These cities account for around three-quarters of our population so, including regional and rural areas (which may have an even higher proportion of homes in close proximity to bush), the number of vulnerable properties probably exceeds 2 million. This research was restricted to bushland, yet grasslands also pose a risk, as the 2005 Wangary fires on the Eyre Peninsula amply illustrated. If we could factor grass fires into the equation, the number of properties at risk would increase still further.

Studies like these make it quite clear where the dangerous locations are. Our urban fringes, with a high density of people combined with large expanses of bush, are the worst possible combination. The concrete jungle, quite simply, is much safer.

But, despite our increasingly urbanised population, we can't all live in cities and, indeed, not all of us want to. Rural industries, such as farming and forestry, remain essential services that need people to live in the country. Urban centres have their own collection of risks, perils and horrors. Many of us might be forced to work in cities but would rather live in the country. The so-called 'tree-changers' explicitly move to the more vegetated perimeter of our cities in order to live in a more natural, greener environment. Many people are willing to accept the risk of a bushfire in return for the benefits of living in a beautiful environment, something researchers call 'the Eden overhead'.[2]

This desire to live in a more natural environment also restricts our ability to control fuel. As one resident of a similar forested area, the central western USA, said:

> You know, when you live out here, it's so nice to be around the trees; that is the whole point of being here. And that is why everybody ... they build in the trees, don't want to knock the trees down.[3]

There is little point in living in a fringe area if all the vegetation has to be removed to reduce fuel loads. The proliferation of smaller blocks on the fringe also reduces people's ability to maintain a cleared, defendable space around their homes. Indeed, while there is plenty of evidence that removing undergrowth and leaf litter from around the home significantly reduces the risk of fire, the benefits of removing trees is less clear.

There is no doubt that fires need fuel, and vegetation, including trees, provides that fuel. Ipso facto, the more trees you have, the more likely you are to get burnt. But fires also need oxygen. They thrive on wind. On Saturday, 7 February

2009, over a thousand fires started in the state of Victoria; nine turned into major incidents—and two into disasters. On the same day, the neighbouring state of South Australia experienced just two fires, neither of which was major. Both states had suffered more than a week of temperatures over 35°C. Both states have the same capacity for disastrous fires. Neither state had received much in the way of rain since mid-December. What was different? Was it just luck?

For most of Saturday, Adelaide experienced moderate winds, gusting around 30 kilometres an hour at the airport. It was enough to raise a bit of dust and blow loose paper around, but nothing extreme. At Tullamarine Airport on the northern outskirts of Melbourne, wind speeds of more than 50 kilometres an hour were being recorded—near-gale conditions that make walking difficult and set whole trees in motion. At Kilmore Gap in Wallan, near the start of the worst fire, a wind gust of 93 kilometres per hour was recorded by the Bureau of Meteorology weather station. Winds of these strengths are storm force—they damage buildings, uproot trees and are rarely experienced far from the sea. Winds of 118 kilometres per hour are classified as a hurricane. Judging from the localised wind damage visible later in the week, we can assume that the winds had been even stronger in some areas of the fire zone.

It was wind that drove the disastrous conditions of Black Saturday, not just heat and drought. Oxygen is a potent accelerant. Strong northerly winds and then the sudden sustained blast of wind from the south produced a fire much fiercer and hotter than anything that might burn on a calm day. Wind blew off roofs, flung burning embers kilometres ahead of the fire front and swept the flames 100 metres into the air. On the cleared edge of the forest, the wind was so fierce that those 100-metre-high flames were occasionally blown flat along the ground, reaching out to touch houses that should, under most conditions, have been quite safe from direct flame contact. It is also possible that the wind kept the hot air close to the ground, blowing across the burning areas instead of releasing it upwards as would happen in calmer conditions.

Conditions like these have been described before on the edge of a forest. On the coast near Apollo Bay in 1851, foresters were driven from their small logging camp into the very ocean itself.

> Having observed that the forest was fired in the distance, they took the precaution on the morning of the Thursday, of burning a girdle of timber around the Bay. In the afternoon, the wind shifting, three fires united in the flames, and came down on the settlers with resistless fury, annihilating everything in their way. Mr. Fisher, C. and another fled to the sea, and stood up to their shoulders in the surf near a reef of rock, for three hours. The flames, though coming as he states, from the forest three quarters of a mile distant, scorched them as they stood in the water and drove on out to sea![4]

Stories like this have seemed far-fetched in the past. Fire does not normally behave this way. Under most conditions, a cleared zone of 50 to 80 metres is enough to protect a house from direct flame contact. But we have very little knowledge about what happens under extreme conditions. We can't simulate or predict these once-in-a-century events or the diversity of conditions they occur in. We can only study them once they have passed. There do seem to be particular dangers on the forest edge that perhaps don't apply within the forest itself.

I live in a gully, surrounded by forest. The wind blows

overhead, thrashing the canopy of the trees, but it rarely gets very windy on the ground. My neighbours live on the top of the hill. In windy weather, deckchairs and blankets go flying from their verandah, scattering themselves throughout the garden. On Black Saturday, my neighbours' house was scattered with ashes and burning bark. In my vegetated gully, I saw not a single ember. Trees may be flammable, but they are also a wind break, shielding the area behind them from the direct effects of wind, and the incendiaries it carries.

Planting or maintaining tree breaks may seem counterintuitive, but they can protect from ember attack, particularly if they are wide enough to deflect the flow of wind over the top of a house. Narrow, solid walls that block the wind often cause air turbulence behind the barrier, which may create more problems than the wall was meant to solve. Similar turbulence may be caused by a large clearing in the middle of a forest. It is not uncommon for townships that have been particularly badly affected by fires to be nestled into valleys and gullies, surrounded by vegetated hills—Mount Macedon, Strathewen and Marysville are all examples of this topography. The impact of wind turbulence on these locations during a fire is difficult to predict. Many firefighters believe that the sudden inflow of oxygen created by a forest clearing sucks the fire in, creating a sudden rush of combustion. Whether this is true or not remains to be proven, but it certainly provides some pause for thought. The relationship between trees and fire is a complex one, and simply removing trees, even if this were either possible or desirable, is not necessarily going to provide the security we all crave.

TO BURN OR NOT TO BURN

The Acheron Way near Narbethong contains some of the most beautiful forest scenery in Australia. As you approach from the Yarra Valley, the road gradually rises from the open, drier woodlands of the valley into the wetter mountain forests. The trees straighten and lengthen as you round every bend and the ground beneath them erupts in a delicate riot of tree ferns and bracken. Eventually the roadside is crowded with the smooth trunks of mountain ash, rising up on each side, like a colonnade for giants. The lush serenity of this river valley could not seem further from its namesake: the ancient river of Acheron in Greek mythology carried mortal souls into hell.

For most of its lifetime, this forest has been wet, moist and hard to burn, even if you try. The gullies down near the Acheron River are home to many rainforest species. Even when the forests surrounding them have burnt, in the periodic, hot, fierce fires these forests are adapted to, these cool damp refuges tend to remain untouched. They are now home to some of the oldest and tallest surviving mountain ash trees.

Today, the once-uniform green that blanketed these hills is gone. Great swathes of black spread across the landscape. The cathedral-like canopy of the forest has disappeared and sunlight falls unfiltered to the bare earth. The massive trunks of the mountain ash stand silent and dead in the ruins of their forest. The fires of hell have indeed changed the landscape of the Acheron to one more befitting its namesake.

From the air, the great tracts of crown fires are revealed in a characteristic crazed pattern of grey created by the dead trees and their shadows, black on black. In the fiercest fires, not a twig or leaf smaller than a finger is left—the entire canopy is burnt back to the stumps of the larger branches. In nature, in this particular type of forest, such fires are rare, catastrophic, but entirely natural. A great many plants and animals die in such fires; not all will have found shelter in the moist gullies that did not burn. Fires are the destructive part of the cycle of regeneration and renewal for this forest. These forests burnt before, in 1939. They will burn again. Without the fires, many species, including the mountain ash itself, would become extinct and the nature of the vegetation and ecology would change forever.[5]

Over two centuries of European settlement, we have halved the area of our forests. Forests now cover just 5 per cent of the country. The remaining forests are often located in dense, inaccessible realms, unwanted for farming and agriculture—the south-west wilderness areas of Tasmania, the alpine parks of New South Wales and Victoria and the Daintree in Queensland are classic examples. Protecting the natural diversity of these forests is essential for the preservation of a great many distinctive plants and animals. Yet how do we reconcile our need to protect these great natural assets with our need to protect ourselves from fire? And how do we protect the small isolated populations of now critically endangered animals from the negative effects of fire, while still allowing the positive benefits of fire to regenerate the bush?

POSSUMS IN POCKETS

The Leadbeater's possum is an exceedingly rare species. It is so rare that, before the 1960s, people thought it was extinct. The species was known only from a handful of fossils and a couple of musty specimens tucked away in the basement of the Melbourne museum. Some of the forests where these specimens had been found had been cleared. There was no reason to think they might be found elsewhere.[6]

But, in 1961, a living Leadbeater's possum was spotted in the mountain ash forests not far from Narbethong. Although some individuals have been recorded historically in other areas, the bulk of the population seems to be restricted to the mountain ash forests of the central highlands of Victoria. Their secretive habits and mountainous habitat makes them difficult to count, but scientists estimate that there are only around 4000 animals left in the wild.

The forests where these possums live were burnt in the 1939 fire and the resulting regrowth has created an ideal habitat for the possums. The large trees killed by the fires provide abundant nesting hollows, without which this species, like many other Australian birds and mammals, cannot survive. The regrowth forest contains large numbers of the wattles that the possums feed on.

As the forest ages, the old dead trees tend to fall down, the wattle understorey is shaded out and Leadbeater's possums move on, presumably to other areas of forest in a fire-induced regrowth phase of their life history. Leadbeater's possums need fire to create a mosaic of different forest types within the mountain ash, so that they can always find a pocket to suit them. With much of the forest gone, and with such large areas burnt in 1939, the possums may now be restricted to such a small pocket of forest that fire could be more of a foe than a friend.[7]

In 2009, more than half of the known area of Leadbeater's possum habitat was burnt out, affecting two of only three locations where the species is known to live and breed. The fires in these forests were so fierce that it is unlikely that many individual animals would have survived. Recovery will rely on surviving animals in nearby forests repopulating other areas. There is no guarantee that such a rare animal, with such particular, and ever-shrinking, habitat needs, will be able to recover at all.

The relationship between fire and conservation is contentious. There is no doubt that fire is a necessary and natural part of many Australian ecosystems and that many species depend on it to create the environment they need. Exactly how much fire, and when, is much more difficult to ascertain, particularly when some endangered species are already under stress from other factors.

The way in which we manage these ecosystems is crucial to the survival of the species dependent on them. Managing fire, particularly in valuable timber forests, has been historically a contentious issue.

INTO THE FORESTS

Forestry departments were established in Australia in the early 1900s. Victoria's forestry department was established by 1908. The New South Wales *Forestry Act* was enacted in 1916. Western Australia followed shortly after in 1918. Even South Australia, with limited natural forestry resources itself, was a leader in forestry, opening one of the first forestry schools at the University of Adelaide.

Forestry had emerged as a profession at the height of the British empire with most of its newly qualified professional foresters being trained in either the United Kingdom or France. The French forestry school at Nancy, L'École Nationale des Eaux et Forêts, stressed the highly centralised and hierarchical model of forestry characteristic of French public service and education. The English forestry schools were dominated by German foresters, who stressed a rationalist model of economic and scientific rigour in the management of forests.

Foresters graduating from these programs aspired to uniform and consistent forests under a national (or even empire-wide) regime that would maximise consistent and predictable yields.[8]

Ecologically, these models were based on a concept of forests as maturing to a stable state. Forests began as seedlings, grew and developed before reaching maturity, at which point they remained stable, uniform, self-sustaining ecosystems. Trees within these forests might be harvested and new ones grown to take their place in the mature forest, but the forest essentially remained the same.

Imposing these ideals onto the Australian landscape must have been an astonishing struggle. Forestry was a means of imposing order and productivity onto a disordered and unproductive natural environment—but what a recalcitrant environment the European-trained foresters found themselves in! Not only did the forests themselves resist this model of neat, self-sustaining, mature forests, but the young professional foresters also struggled to impose control over the rugged woodcutters and bushmen who dominated the industry. These men, like the surrounding landholders, came from a very different background—one deeply steeped in individualism, contempt for authority (and learning) and with ingrained local knowledge. They had absorbed the Aboriginal practice of fire-stick burning and adapted it to their own purposes. Their resistance to national models of forestry was also supported by the division between Australian states. Practices that were deemed to be practical and suited to local conditions were often maintained in defiance of approaches derived from scientific or theoretical regimes.[9] The use of fire in and around the forests was a perfect example.

Unlike northern hemisphere forests, Australian ecosystems rarely have a stable, mature form. Adapted to suit long cycles of drought and periodic, patchy burning, Australian forests are best described as a mosaic of different aged and structured ecosystems. Within the same forest, one area might be dominated by short-lived regrowth wattles, producing a dense, temporary understorey, while in an adjacent patch, rainforest species are gradually overwhelming the dry-adapted eucalypts. These cycles occur across time and space, maintaining a patchwork of different habitats, supporting a diversity of species that move within the forest to find the conditions that suit them. It is fire that creates and maintains this patchwork, by periodically burning small sections of forest and leaving other areas untouched.[10]

Still, forestry men worked hard to impose strict hierarchy, bureaucracy and control over the forests and bushmen of Australia and strict fire suppression formed a part of that process. Just as it was in the French forests of Nancy, fire in Australia was seen as a destroyer of valuable timber rather than a natural part of a healthy ecosystem. Fire was an uncontrollable and chaotic element that needed to be eliminated. While in France most people probably agreed with this view, the same could not be said of Australians.

In its first report, the Victorian Forests Commission noted:

> Many people have no compunction in setting the forest ablaze, and who for small immediate gain, can watch without pang the immediate destruction in flame and smoke of the timber that might provide their sons with employment and wealth.[11]

Supported by the devastation caused by fires in the 1800s, government policy across the southern states soon moved towards fire suppression. The success of these strategies was limited. Firstly, not everyone was as keen on fire suppression as foresters and government officials. Even into the early 20th century, farmers were still illegally clearing land with fires at the height of summer. One Otway Ranges resident recalls how a settler would drive a candle into the ground surrounded by kerosene-soaked bagging to act as a slow-burning wick. A few hours later, the culprit could be innocently sitting in a public place well away from the scene when the first report of a fire on their property came through.[12]

Secondly, the result of long-term fire suppression in some cases increased the risk posed by fires. By the 1930s, fuel loads in many Victorian forests were high and, when disaster struck in 1939, many blamed the Forests Commission. In the words of George Purvis, a Gippsland grazier, in 1939:

> Nowadays, if we want a fire we nick out in the dark, light it, and let it go. We are afraid to tell even our next door neighbours because the Forests Commission is so definitely opposed to fires anywhere, that we are afraid to admit that we have anything to do with them.[13]

Similarly, fuel loads in many West Australian forests in the 1950s were regarded as being dangerously high. The system of carving the forest up into blocks with small tracks was clearly inadequate as a means of preventing large fires from moving through the forest. The fires near Dwellingup in Western Australia illustrated the disastrous potential for fuel build-up.

From the 1950s onwards, policies gradually moved towards regular rotational fuel reduction burning across Australia. Where applied, this appeared to successfully reduce fuel loads and the intensity and severity of fires in the forests, while simultaneously improving some ecological values.[14]

Fuel reduction burning is not a particularly easy option. It is time-consuming and resource intensive. It always carries the risk of fires getting out of control. Somewhat surprisingly, the smoke often annoys the very residents the burns are intended to protect. While governments are only too happy to open their wallets and cheque books in response to a fire disaster, they are less forthcoming with everyday funding for mundane fire protection activities, even though these may be considerably cheaper and substantially reduce the losses caused by fires. Everyone loves a big firefighting helicopter or a shiny, new, red truck. A man with a drip torch wandering through the forest is not quite so headline-grabbing.

Today, the various parks and forestry authorities combine their efforts (to a greater or lesser degree) with their colleagues in various fire agencies. Over time the former forestry departments have been amalgamated into bigger and bigger super-departments encompassing the traditional portfolios of lands, conservation, mining, wildlife, fisheries, water and parks. These departments each have their own histories, traditions and approaches, and are riddled with internecine struggles. Some areas are natural allies; other are sworn enemies, yet their political masters throw them all in together under the mistaken belief that one public servant is much the same as the next one. Their departmental fortunes and relative powers have waxed and waned over the years. A once-powerful lands department might all but disappear as the supply of Crown land is gradually appropriated by other specialist departments. A poor public servant cousin charged with caring for national parks might slowly and steadily transform into a dominant partner, in line with its growing responsibilities.[15] The ghosts of old allegiances and heritages linger in the corridors of bureaucracy. Fire suppression and prevention, even in the most modern conservation department, often still retain the hallmarks of old forest protection practices, for all their eco-friendly outer garments.

Public land may only account for 23 per cent of the Australian continent but this land contains three-quarters of our high bushfire-risk forests. Clearly, the firefighting capabilities and approaches of the various government departments and agencies responsible for the care of public lands, and particularly the forests, are vitally important to our overall approach to bushfires.

WHEN TO BURN

In the absence of clear information about fire regimes, some land managers take a precautionary approach—if in doubt, don't. If there is any concern about the impact of burning on rare plants or animals, they don't burn. David Ward, a former research scientist from the then Conservation and Land Management Department in Western Australia, argues that we should be adopting the reverse strategy. Given that there is clear evidence that burning was more frequent in many ecosystems in the past, he maintains that the safest option is to burn, unless we are sure there is good reason not to.[16]

Burning for ecological reasons is highly contentious and disputed. Most ecologists and conservation biologists would probably agree that it is a desirable course of action, provided the burning regime is similar to that which occurred in the past. It is probable that many controlled burns are bigger and more uniform than they ideally should be. The old ideologies of the forests still drift through the corridors of the public service, ready to do battle with the equally dogmatic beliefs of conservation ideology. We need more information on what fire regimes occurred in the past, before European settlement, perhaps even before Aboriginal occupation. The information is often available—in the historical records and in the environment itself. We don't always have the resources needed to pull it all together.

Other conservation activists are not convinced that burning the bush is a good idea. The Australian conservation movement, like many other aspects of our lives, is heavily influenced by concepts developed overseas. The notion of undisturbed, natural, old-growth forests is a pervasive and attractive one, which many people find difficult to relinquish. Australia's ecosystems are far more complex and cyclical than that. Disturbance is a natural part of our cycle, and old is just a step on the path to being new again.

'Undisturbed' implies being without human interference—and humans certainly have a remarkable capacity to stuff up the things they fiddle with. But this idea does assume that leaving things alone is enough, that our broader activities have no effect. The Australian landscape has been modified and altered by humans for many thousands of years. It has probably only been modified a little bit for most of that time but it has been modified a lot in the last 200 years. Leaving things alone will not return them to 'normal' or to some predetermined stable state, because that is not how nature works, particularly not in Australia.

What is natural is for species to become extinct, for invading species to take over and dominate areas they never occurred in before, for forests to retreat, expand, even disappear, with changing conditions. We can opt for minimal management but the end result could just as well be described as 'neglected' as 'undisturbed'.

Natural, undisturbed vegetation does not always have to have a dense, thick tangle of undergrowth. These conditions suit some species but not others. Some species require thick, woody undergrowth; others only thrive in an open forest with a clear understorey. What is important in conservation is to have a diversity of habitats over a large area to suit the particular needs of different species at different times.

There does not have to be a conflict between fire management and conservation management. Often the two systems can collaborate very successfully. In California, Lassen County Fire Safe Council had very little success in attracting property owners to sign up for their fire protection programs but, when they redesigned their program to focus on 'forest health' instead, enrolments went up from 20 to 80 per cent. From a fire point of view, the authorities wanted to reduce the amount of undergrowth. They noticed that most of the undergrowth consisted of shrubby weed species that had invaded disturbed areas. Removing this undergrowth re-created the open, treed meadows early settlers had described, promoting healthier forest ecosystems and reducing fuel loads.

Hotspots Fire Project

Currawinya, in the high country of northern New South Wales, has seen a lot of changes over the last 200 years. Like much of the area, the valuable red cedar and hoop pine forests were all but wiped out by logging. Years of grazing by cattle continued to change the landscape, increasing weeds, compacting the soils and promoting erosion in the steep river gullies and ravines.

Thirty years ago, the 2125-hectare property was purchased as a refuge for wildlife, and the shareholders and families who live there. Like many alternative communities, the new owners seemed to have little in common with their traditional farming neighbours. Many farming practices seemed to be the cause of the environmental degradation Currawinya was looking to heal; many of the ecologically sensitive practices of the new owners were seen as anathema to the sound land management practices of the farmers.

Like other properties, Currawinya had a huge weed problem, along with feral pests and erosion issues. But the property also harboured many endangered species, including the brush-tailed rock-wallaby. For many years, fire was regarded as the greatest threat to these species.

In October 2000, a hot, fierce fire burnt out a significant area of the property, destroying the canopy, and along creeklines and the sensitive river frontage. The environmental cost was huge. With a drying climate seeming to make such fires even more probable in future, the residents recognised that fire management needed to be part of their environmental planning.

In 2005, Currawinya hosted a two-day Hotspots Fire Project workshop, run by the Nature Conservation Council in collaboration with the Rural Fire Service. These programs bring neighbouring property owners (both private and agencies) together to learn about managing fire in a sustainable way. The workshops help landowners to identify the different ecological and personal assets they have on their property, and work together to implement different fire regimes to protect them from uncontrolled bushfires.

Today, Currawinya uses fire to control weeds, reduce fuel loads and improve ecological values of the property. Even more importantly, the Currawinya community works with neighbouring farmers and agencies and many people from Currawinya have joined their local fire brigade, taking the best from both traditional farming practices and environmental principles to improve land management, conservation and fire safety across the community as a whole.[17]

No matter what your perspective of the conservation merits or dangers of controlled burning, nearly everyone will agree that removing introduced weeds is a good thing. There is often a lot of noise made about local council regulations preventing private landholders from clearing for fire safety. Local bureaucracies and regulations are often unwieldy, poorly considered and badly implemented. It is not always clear how to implement both fire safety guidelines and conservation guidelines. Sometimes the staff responsible for implementing the regulations may not have much expertise in either area.

But it is all too easy to blame local regulations for preventing the removal of native vegetation when much of the high fire-risk undergrowth is made up of weeds, the removal and control of which is actively supported and encouraged. While some weeds can invade undamaged bush, most become established when the natural forest ecosystem has been disturbed by human activity. Building roads, logging, house construction, gardening and clearing all create gaps in the vegetation and allow weeds into the bush. Once established, they can be hard to remove. A great many weeds increase local fuel loads and increase fire risk.[18]

On my own property, the vast bulk of the undergrowth consisted of blackberries, Spanish heath, introduced pittosporum and Cootamundra wattles. Removing these species has opened up the undergrowth, allowing the kangaroos to graze and keep the grass short. Weed removal has gone a long way to reducing the fuel loads around my home, and particularly removing the high-risk understorey that forms a ladder for fires to climb into the canopy. The same principle would probably be true of any number of other weedy woodland invaders in other states, like lantana in New South Wales and Queensland. Many introduced pastures in Queensland and the Northern Territory—like buffel, para, mission and gamba grass—have dramatically increased fuel loads. Removing and controlling weeds goes a long way to making many ecosystems safer from fire and healthier at the same time, without any hint of conflict between fire and conservation agendas.

Not all weeds are a high fire risk, and many invasive high fire-risk species are native. Pittosporums are often recommended as a fire-resistant species that may actually reduce the fuel load. Perhaps this depends on how old or how dry they are. On my place, they go up like crackers in a fire—and they completely dominate the mid-storey, creating impenetrable thickets that are the perfect ladder for a ground fire to move into the canopy. So I remove them. The variability caused by local conditions may explain why fire authorities have moved away from providing lists of 'fire-resistant' plants for householders. Fire resistance depends too much on time and place, rather than just species.

Other native species, like burgan, are both invasive and highly flammable. Burgan is generally only a problem in areas that either have been cleared or are adjacent to cleared areas. On properties with conservation overlays it is often illegal to remove burgan trees, despite their risks. But there are rarely controls on pruning trees, particularly removing lower branches, and there are rarely controls on managing seedlings. Burgan is highly fire sensitive, so well-managed control burns can prevent its establishment. With a bit of assistance, other larger trees can be encouraged to outgrow and shade out burgan. Managing fire risks like burgan may just require a bit of creative thinking. Good land management takes time, effort and knowledge. All too often fire safety is used as a 'quick fix' excuse for years of neglect and apathy.

Controlled burning is an important tool in the arsenal of all land managers, large and small, public and private. Burning under mild conditions allows fuel to be reduced before the extreme conditions of the peak fire season. In the south, this typically involves autumn or early spring burns, when the vegetation is dry enough to burn but the conditions are not so dry or windy that fires can't be controlled. Autumn burns have the advantage of promoting biodiversity and often have the best outcome ecologically. Spring burns can be beneficial for reducing the growth of spring grass before it cures in summer and provides the lowest fuel conditions, just before the danger period.

In the north, the season of burning also makes a big difference in the ecological impact of the fires. Areas burnt in winter tend to favour shrubs rather than spinifex grasses. Winter burns are patchier, cooler and cause less damage from soil erosion and to animal communities than the hotter, larger summer fires.[19] The best time of year for burning depends very much on the local conditions, the ecosystem and which species you wish to favour.

In the Northern Territory, there are very few areas of land that aren't burnt each year and almost none that remain unburnt after three years. This regularity of burning undoubtedly contributes to the 'coolness' of the fires when they do occur and their lack of severity. Where vegetation does manage to build up over a few seasons, resulting fires can be both hot and fierce, and much more dangerous.

The words 'control' and 'burn' are not natural bedfellows. Even with the best planning in the world, burn-offs can get out of control. The window of opportunity for burning, relative to the area of land that needs burning, is also small. Burning too close to the fire season increases the risk of the fire getting out of control. Burning too close to the wet season often means the fire won't burn at all or doesn't reduce enough fuel. Despite the fact that such burns are done almost entirely to protect human life and property, many people grumble and complain endlessly about the smoke and inconvenience of the burn-offs, although these complaints, consistent though they are, are never quite as loud as they are about the lack of controlled burning when a bushfire strikes. The authorities who are responsible for the largest tracts of forested land are all too readily damned if they do and damned if they don't.

ROADSIDE VEGETATION

Another conflict to emerge between environmental and fire safety practices arises from the issues of roadside vegetation. From an environmental point of view, roadside vegetation forms essential corridors through an increasingly hostile environment for our plants and animals, allowing them to disperse and move between otherwise isolated fragments of their habitat.[20] Small, isolated populations are much more vulnerable to extinction and less able to recover from damage.

Cleared land is often a very hostile place for many native species. An open paddock may be more of a barrier to these species than a razor-wire fence. This is why roadside vegetation is so precious. The black bitumen strip may be our means of transport but the tangled mass of bush on each side is a tenuous transport link for a surprising number of native species, birds and mammals alike.

On the negative side, a great deal of the tangled mass of vegetation on roadsides is weeds. Roads are, in themselves, a major disturbance to natural vegetation, so they are inherently prone to invasive species. Removing the weeds from roadsides is likely to significantly reduce fuel loads, while still retaining their conservation value.

Unfortunately, not only wildlife travels along these attractive green verges; they are also corridors for fire. Driving along some of the burnt-out roads of Kinglake, I realised that what I had thought was thick vegetation was actually just a narrow strip concealing the open paddocks behind. Driving along these roads in a fire would have been to drive through a tunnel of flames. It must have seemed that the whole world was on fire, despite the fact that just metres away lay the relative safety of ploughed paddocks.

The safety of roads during fires is a major concern for residents. Many people are obstructed by falling trees, or caught in flames from burning vegetation. The dangers of being on the road during a fire cannot be overestimated. Fire trucks don't travel down roads while they are burning. Nor should anyone else. Cars and roads are a dangerous place to be, even in a grass fire. Clearing every tree along the roadside, even if it were possible, wouldn't necessarily make it safer to drive through flames and smoke. A great many people who die in fires on the road die from collisions or accidents. The only way to make roads safer during a fire is not use them.

12 · THE BUSH BRIGADES

No man is an island, entire of itself; every man is a piece of the continent, a part of the main. If a clod be washed away by the sea, Europe is the less, as well as if a promontory were, as well as if a manor of thy friend's or of thine own were: any man's death diminishes me, because I am involved in mankind, and therefore never send to know for whom the bells tolls; it tolls for thee.

John Donne (1572–1631)[1]

FOR WHOM THE BELL TOLLS

The old school bell at the front gate of Panton Hill Primary School in the outer foothills of Melbourne has rallied its community for decades. Today, its services have been replaced by a piercing electronic chime but, years ago, children swung on its rope to ring their classmates in and out of school, recess and lunch. Every morning, the bell rang five minutes before the school day started, prompting the dawdlers to dash the last stretch of their hike to arrive, puffing and panting, just before the teacher.

The bell has rung in celebrations and commemorations. It has rung in the coronations of kings and queens. It has tolled the end of the world wars and marked the passing of young men who never returned. And it has called the community to arms for local battles. Whatever the emergency, whatever the crisis, the bell has called the men from mines, farms and orchards, the women from their houses, shops and gardens, rallying the community in an hour of need.

Pubs on fire, lost children, mining accidents—all these disasters have been heralded by the bell. In the summer holidays, everyone knew what it meant to hear the school bell ring with strident urgency in the late afternoon heat. Grabbing buckets, hoes, rakes, sacks and axes, the men leapt on horses and carts, bicycles and old trucks and roared in a cloud of dust to the fire.

There was little coordination of these firefighting efforts and even less equipment. Water was a precious commodity rarely available on the fire ground. Urban brigades might have pumps and water, brass-buttoned uniforms and helmets, but in the country there was only raw courage and bare hands standing between the fire and the farms. Men and women stood at the edge of the fire, beating it out with whatever came to hand, wet sacks if they were lucky, tree branches if they were not. In Panton Hill, as in small country communities right across the continent, fighting bushfires was just a fact of life on the land.

THE CITY AND THE COUNTRY

The fire brigades in the major cities were originally organised by insurance companies, keen to reduce their own liability. By the late 1800s, most states had passed a *Fire Brigades Act*, offering varying levels of coordination of the urban firefighting efforts.[2] Many of these early brigades were manned by volunteers and often coordinated by local policemen. As the urban centres expanded, the need for paid full-time firefighters grew and the volunteers were often unceremoniously dumped.[3] The small teams of paid firefighters fought an uphill battle against fires and public derision. Slow turn-outs, lack of organisation and dubious equipment often meant bucket chains of locals were seen to be more effective on the fire ground. Smart uniforms and gold medals in firemen's games clearly appealed to those in charge of the services but did little to inspire public confidence.

Urban firefighting is much the same the world over and models for fire brigades, equipment and organisations based on British systems were readily adapted to Australian towns and cities. But Australia presented a unique challenge to firefighting traditions from overseas. Models of urban firefighting offered little in the face of that distinctive feature of the Australian bush—the bushfire.

In the rural areas, fires were a seasonal threat, rather than the regular threat that they posed in the city, and required a large number of firefighters for a relatively short period of intense activity. There were fewer assets for insurance companies to protect in the country and less money available to support paid firefighters. It was simply not economic, or even feasible, to employ enough firefighters to battle against really large fires. Many regional brigades, such as those started in the goldfield towns of Victoria, soon recalled their volunteers.[4]

In rural areas, firefighting remained the responsibility of the local community. Many communities, particularly in the south-east, started their own 'bush fire brigades', fiercely independent from the state-organised metropolitan brigades in the cities or the so-called country or rural brigades that operated out of the larger towns.

THE BERRIGAN BUSHFIRE BRIGADE

The first of these bush brigades was founded on 26 November 1901 in the tiny Riverina town of Berrigan, just north of the Murray River. The Riverina, like much of southern Australia, was at the tail end of the decade-long Federation drought. The country was so dry that in places it was said to be possible to cross the Murray itself without getting your feet wet. The Riverina had been particularly hard hit. Production had been slashed to a fraction of earlier years and many struggling farmers were forced off their land. As if drought weren't enough to contend with, aridity brought his psychopathic sister to the party and fires raged across the country.

In January 1892, fires broke out almost simultaneously across the Riverina. Thousands of hectares of wheat and pasture were burnt, destroying the livelihood of the local farmers for the next year. Some blamed disgruntled swagmen. Whatever the cause, the swagmen soon had to jostle for space on the tracks with thousands of burnt and hungry sheep in search of food and shelter.[5]

Three years later, and fires again burnt through Berrigan, burning north along the stock route from the river town of Tocumwal, again destroying several local farms. The final straw must have come in autumn, when the people of Berrigan should have been safe from their seasonal curse. Late at night on 13 April 1897, a fire:

> started at Drohan's barber's shop, and quickly extended to Kennedy's tailor's shop and Williamson's plumber's shop, adjoining Garrard's general store, containing a very valuable stock, which was saved. The flames, however, blew across the road and destroyed Simmons's bakery, and Malan's fruit shop, in which

> was a school and a library, Cameron's saddlery and Roberts' hairdressing saloon, and Trask and Whitty's (solicitors) offices.[6]

No-one in Berrigan needed to be reminded of their vulnerability to fire by the disastrous summer of 1897–98, which began with six people dying in fires around Hobart in December, and culminated in Gippsland's Red Tuesday, where twelve people lost their lives. The first of Australia's bush brigades was apparently underway by 1896, and formally recognised by 1901.

Berrigan's bush brigade was the first of many to spontaneously spring up around the country, driven by the community's need to protect itself when no-one else could or would.

BRINGING THE BATTLE HOME

The bush brigades were often manned by farmers, who provided their own equipment, including water carriers. In more isolated areas, some fires might be contained within the confines of one or two properties, so naturally responsibility for such fires fell on the landowners and their neighbours. Large outback stations might have enough manpower to constitute a firefighting force in their own right. In more densely settled areas, bush brigades centred on small townships.

The bush brigades operated with various levels of organisation, some meeting only once a year to lay out their pooled private resources to see who had what. Some used a school shed or community hall for storage and as a rallying point. Others met more regularly and some even had their own sheds or trucks. All operated on the basis of a shared sense of community responsibility, a 'one for all and all for one' principle, recognising that a fire on your neighbour's land today could be on yours tomorrow.

While the urban brigades were frequently manned, either voluntarily or paid, by people who wanted to be firefighters, the bush brigades were quite different. Members joined these brigades because the fires had to be fought and because there was no-one else to fight them. When the fires were over, they went back to their farms, trades and other occupations. This fundamental difference in nature has reverberated through the decades, even where both types of brigades have been amalgamated under the one organisation.

Like many rural communities across Australia, the people of Panton Hill knew they needed a fire brigade. In early 1914, after a hot, dry summer, a public meeting was held in the hall to 'consider the advisability of forming a Bush Fire Brigade at Panton Hill'.[7] They needed somewhere central to store equipment, to have some system of organisation to plan and to train for fires, and to lobby for support. But those who volunteer to fight fires also volunteer for other duties—nearly one-third of Victorian volunteer firefighters enlisted in World War I,[8] just as they did in all the other states. The war took away the young men of the towns and country and diverted the attention of the community to a threat on distant shores.

But bushfires don't stop for the wars of men. The inland fires that raged across central Queensland in 1918 arguably caused more devastation to local economies and communities than the war in Europe, for all its death and destruction. Young or old, whoever remained home answered the fire call. Age did not deter 76-year-old Fred Bartlett from fighting a bushfire that threatened his home in Panton Hill in 1914. He contracted pneumonia for his efforts and died soon after, an unmarked casualty of the war at home.[9]

When the men returned from the war, their experiences

had taught them the importance of discipline and coordination in extreme conditions. It also exposed them to the dangers of poor leadership and bureaucracy. They took these experiences home with them to the fire ground.

The widespread Black Thursday fires of 1926 across the eastern states cost lives in Victoria, Queensland and New South Wales. Various governments responded by overhauling their fire legislation, providing greater statutory support for the fire brigades, although not providing much in the way of practical support on the ground. Forest services began to put more resources into fire control, recognising the significant risks and costs associated with fire in the publicly owned forests. The standard equipment for a rural fire brigade at the time was an axe, rakes and knapsack sprayers. A rural brigade with a truck and a hand-operated pump was well equipped indeed. Generally, the bush fire brigades relied upon farmers to provide tractors and other agricultural machinery that could be adapted for firefighting.[10]

It always takes a major tragedy to really get people organised. The fires of 1939 provided the impetus for many brigades to form around the country. On 14 January 1939, the day after Black Friday, Panton Hill held the inaugural meeting of the Panton Hill Fire Brigade, joining nearly 400 bush fire brigades who took responsibility for local fire protection into their own hands, with assistance, not from the metropolitan or 'country' firefighting authorities but from the Forests Commission.[11] Their equipment, such as it was, was housed at the school, close to the bell, which still sounded the alarm and roused the community to action.

Bushfire services state-by-state

Australian Capital Territory

The ACT Bushfire Service is administered by the Emergency Services Bureau, along with the ambulance, emergency services and fire brigade. The Bushfire Service is divided into two departmental brigades—forests and parks—and also has nine volunteer brigades across the territory.

New South Wales

The Rural Fire Service responds to fires across 90 per cent of the state. Around 2500 brigades are made up of 70 000 volunteers and nearly 500 staff. Most brigades were originally organised by local councils or communities but are now controlled by the Rural Fire Service.

Northern Territory

Bushfires are overseen by the Commissioner of Police through the Fire and Rescue Service, with coordination and planning provided by the Bushfire Council. The service has around 150 full-time firefighters and 50 part-time firefighters with an additional 250 volunteers.

Queensland

The Fire and Rescue Service currently comes under the control of the Department of Emergency Services and combines both urban and rural firefighters. The urban stations are primarily crewed by paid staff, while volunteer firefighters run most of the state's 1600 rural fire brigades.

South Australia

The Country Fire Service has over 400 brigades across the state, including brigades formed with Forestry South Australia and the South Australian National Parks and Wildlife Service. It employs 70 people and has over 16 000 volunteers.

Tasmania

Individual fire brigades were originally managed by their own boards but were combined into the Tasmanian Fire Service in 1979. Tasmania has 240 brigades, with four full-time staffed brigades in the major centres. The remaining brigades are primarily operated by 5000 volunteers.

Western Australia

In 1999, the Bushfire Service was amalgamated with all the other emergency response organisations under the Fire and Emergency Services Authority of Western Australia. Fire and Rescue combines both urban and rural firefighting responsibilities and has over 800 paid firefighters and 2500 volunteer firefighters across 144 bush fire brigades.

Victoria

The Country Fire Authority has overseen fire services in rural and regional centres across the state since 1945. It has nearly 300 paid firefighters, and 66 000 volunteers across 1200 rural and urban brigades. The Department of Sustainability and Environment is responsible for fires on public land.

SUPPRESSION: TRUCKS, TANKS AND CHOPPERS

It has taken humans a remarkably long time to come to terms with the intrinsic nature of fire in the landscape. For a long period, fire was no doubt equated with some kind of supernatural force—a punishment from the gods or ancestors for wrongdoing. In more recent times, humans have fought back against nature.

Fire suppression has long been the dominant model in both Europe and America. Fires were seen as dangerous and damaging, both to the forests and to people, and all efforts were directed at suppressing them as soon as they were detected, wherever they were detected.

It is no surprise that in the USA suppression is taken to an extreme form. The American fire historian Stephen Pyne describes it as 'the muscle-bound paramilitary response'.[12] This approach is probably best illustrated by the role of the 'smoke jumpers'. Smoke jumpers, armed with a backpack of high-tech goodies, a fire survival tent and small arsenal of explosives, are dropped in to remote forest areas where a lightning strike has lit a fire to put it out before it gets established. Bigger fires demand more smoke jumpers. Apart from their hefty 36-kilogram personal kits, a series of boxes are dropped after the smoke jumpers with extra supplies, machinery and equipment. There's enough stuff to set up base camp near the fire and provide enough resources to carve helipads out of the forest and establish a control line.[13] It is an astonishing thought: that anyone would parachute a firefighter into the middle of a rugged and remote wilderness area to put out a fire, with little more to protect them than a so-called fire tent. Sure, the fire tents are specially designed with laminated reflective fabric—but they are still just tents. They still start to disintegrate at 300°C. A ground fire is classified as slow and smouldering at 500°C; crown fires regularly reach temperatures of 1500°C. No tent is going to protect you from that.

America's history is littered with wildfire tragedies. From the 1870s to the 1920s, numerous fires burnt great swathes of the country, each with death tolls in the hundreds or thousands. The Peshtigo fires, which killed over 1000 people, were followed in 1910 by the Big Blowup, where 85 died fighting fires. The loss of essential timber caused by these fires, as well as the loss of life, finally prompted action and gave rise to active fire suppression activities by the Forest Service. By the late 1930s, the Forest Service adopted a '10 a.m. rule'—all fires were to be brought under control by 10 a.m. the next day.[14]

Australia has never had either the manpower or the resources to combat our much larger fires on quite the same scale as the Americans. Our fires are bigger and our population smaller. Nor, it seems, has there been the inclination.[15] From the very earliest days of European colonisation, white Australians have recognised that burning plays an important part in the Australian ecosystem. While total fire suppression was a popular policy in the 1950s, it was never particularly successful. Fire has always been an important tool both in land management and in Australian firefighting. Perhaps it's not surprising that while Americans were perfecting water-bombers to suppress fires, Australians were developing fire-bombers to back-burn or control burn during low-risk conditions.

Australia cannot suppress all its fires. Our approach has, by necessity, had to be much more pragmatic and inventive. We have had our share of failures but, in all, our successes are significant and the lessons we have learnt are worth remembering. Bushfires may rate as Australia's worst natural disaster but, given our huge exposure, deaths from bushfires remain relatively small, particularly given the large number of people who live within the fire zone.

Every death from fire is a tragedy and worth working to prevent but we should not forget that, by and large, we do cope well with fire in this country. We should not let ourselves be overwhelmed by the tragedies and by what went wrong to the extent of ignoring all the many things that went right. We spend a lot of time worrying about what went wrong. We can learn just as much, if not more, from what went right and work out how we can do that even better.

Australians have a proud history of facing fires, fighting fires and living in a landscape dominated by fire. Historically, individuals and communities have faced the threat together, working side by side to defend themselves, their families and their neighbours. We have not, in the past, looked to governments and authorities to save us. We have expected them to coordinate, organise and support our efforts. Maybe we can and should expect them to do that a bit better.

Our bushfire response is overwhelmingly based on a grassroots local community movement. Across Australia, more than 200 000 volunteers make up the backbone of our firefighting services. Not paid firefighters, not specialists, not people who have made this their occupation, but ordinary men and women—accountants, farmers, truck-drivers, builders, academics, electricians, shopkeepers, managers, winemakers and artists, even the odd politician or two. These are ordinary people who take responsibility for actively protecting themselves and their communities.

Not all of us are suited or able to meet the challenges and obligations of being a volunteer firefighter. But we can all do a little bit better to help ourselves and each other, to make sure we are not adding to the burden on others. Many of us claim that we don't have the time to volunteer or to prepare. Being a volunteer firefighter does take up a lot of time. So does preparing our homes and ourselves for bushfires. If we don't have the time to make sure we can live safely in the bush, then perhaps we need to think about living somewhere else.

The key to our safety from fire lies, as it always has, not in fire trucks and helicopters, not in text messages and evacuation orders, but in ourselves.

13 · BE PREPARED

SHOULD I STAY OR SHOULD I GO?

Moving to the outskirts of the city was the best decision we ever made. Living in the inner city in a cute little heritage worker's cottage certainly had its advantages. I rarely used the car, could walk to most shops and was spoilt for choice of public transport. Work was a pleasant walk or ride through parklands. Dining out was a multicultural joy and all the major facilities were within easy striking distance.

But after a few years of being on intimate acquaintance with our neighbour's plumbing and yet barely being on speaking terms with anyone in the street, the lure of the 'tree change' grew. Cheap land, compared with spiralling inner-city prices, was also highly attractive and we opted to exchange our inner-city lifestyle for a semi-rural idyll.

Like most southern Australians, we knew very well that the area we were moving into was at risk of bushfires but this risk wasn't at the forefront of our minds when we purchased our new home. We debated the relative merits of hilltop properties with northern views of the ranges but eventually selected a house in a south-facing gully, not because it was safer from fires but for privacy. The house appealed to us (although not to many others) for its solid and classic 1970s architecture, not because its concrete slab and lack of eaves made it easier to defend from fire.

Our new home required a lot of work. The creepers and ivy that covered the house had to be removed for basic maintenance. Gravel paths and drains had to be installed all around the house to improve drainage and stop the flooding that the house was prone to. An abundance of weeds and shrubs had to be cleared from around the house. The dam had to be repaired so that it could fill with water instead of letting it gush down the gully from a hole in its bank. All these decisions were made primarily for other reasons—practicality, safety, conservation, aesthetics—but behind all of them also lurked a vague concern about fire safety.

Our knowledge of fire preparation was fairly rudimentary when we first moved to the area. Like many others who move on to bush blocks, we worried about the lack of an alternative escape route from the property. What would happen if a tree fell across the driveway during a fire? There would be no way out. We pondered the options of tracks to neighbouring properties. It was many years before the illogicality of an 'escape' route finally sank in.

Just a few days after we moved in, we were visited by the captain of the local fire brigade, who happened to live down the road. His friendliness encouraged my husband to join the local brigade as a volunteer. As Mike's involvement in the fire brigade increased, our understanding of fires improved. We went to the occasional street meeting and, more importantly, read all the pamphlets and brochures on fire safety. We learnt, from firsthand experience, how easily small backburns get beyond one person's control in dry conditions. We learnt how fires that you think are extinguished smoulder on undetected and flare up unexpectedly. We learnt how hard and easy it is to burn different materials, of different dampness and under different conditions. None of this would help in a big fire but it would at least help us to recognise what we didn't need to panic about. We realised that there was no need for an escape route. If you are home when there is a fire in the area, there is no safe way out. You have to be prepared to stay and defend.

LEAVING EARLY

When we first moved here, my eldest daughter was four and my youngest was a slowly expanding bump. It was very soon apparent that in the event of a fire, Mike would be out on a truck. There was no guarantee he would be able to get home to help me if a fire threatened our area. Staying and defending our home on my own in my condition simply wasn't an option.

The official advice was unambiguous. If you do not intend to stay and defend your home in the event of a fire, you should leave your home early in the morning or the night before a high fire-risk day. Identifying a day of high fire-risk is pretty easy. They are hot and often windy, with low humidity, and often occur after a few weeks of hot, dry weather. They are nearly always declared total fire ban days.

The decision to leave early was generally straightforward. When I was still working in the city, my eldest daughter was in child care near work, so it was easy not to be at home on risky days. We had no pets to worry about. Sometimes it was a simple as having dinner in town after work while we waited for things to cool down. Later, as we spent less time in the city and more time based around the local school and kindergarten, the decisions became more difficult. But it was still a relatively straightforward choice. Every total fire ban day, I packed a few valuables in the car and took the kids to the movies, the museum or a shopping centre for the day. Once or twice, I imposed on friends in the suburbs for a visit. It wasn't difficult or complicated. My decision was to leave, and so I left. Not when there was smoke, not when there was a fire nearby, not even when there was a fire further away. I left every high-risk day, every day that was declared a total fire ban.

On average in Victoria there are only 10 total fire ban days a season. In a bad year there maybe as many as 25, in a good year fewer than 5 such days. Although it seems that in recent years the number of total fire bans has increased, on average there are no more declared now than there were in the 1960s.[1] Some people claim that, if they were to leave the area on every high-risk day, they would have to leave their homes from November until May. This is an exaggeration and suggests they have not really given much serious consideration to what conditions are dangerous, and when they need to leave. Not every day of summer is a high fire risk, not all total fire ban days are necessarily that dangerous. Sometimes the day itself isn't so bad but the next day has the potential to be so awful a total fire ban is declared just to reduce the risk that any fires will be going when the bad weather arrives.

FIRE RISK

It should really be quite clear which areas and which locations are at risk from bushfires. The risks are largely physical: proximity to forest, slope, exposure to wind. Given our propensity for self-delusion when it comes to unpleasant things like fire, perhaps it is worth spelling it out, making sure every single property in risk areas is clearly assessed and labelled in relation to its relative vulnerability to fire. Some states and some councils already provide this information. In Queensland, for example, council area maps of fire risk are publicly available on the internet. Such information is often available to councils and local fire authorities but is often difficult and complicated for the general public to access. Such maps also need to be quite detailed, relating to the risks for each individual property. There are so many different smaller-scale projects tackling this need that the ideal solution would be for the different approaches to be assessed nationwide, with the best practices identified, standardised and adopted in a uniform system across the country, while still taking into account particular local needs and conditions.

With local councils already having responsibility for planning and development, local government is the obvious body to be responsible for collating and maintaining this data. But with 677 councils across Australia, all with varying levels of expertise, funding and organisation, some national or state level of management and funding is clearly required. At least then everyone would have a clear, unambiguous assessment of the risk their particular property poses, which then provides people with an opportunity to assess what exactly is required to make their property safer.

Understanding the environment we live in, and how that environment is affected by climate, seasons and weather, is essential to understanding fire risk. The more removed we are from the natural environment, the more difficult it is to assess where dangerous areas are, and when they are at their most dangerous. Once we know how much risk a particular property faces from fire, we then need to keep tabs on when it is most at risk. Basically, that comes down to watching the weather.

WHICH WAY THE WIND BLOWS

We need to pay more attention to the weather. When we work in the city, there are really only two things we're interested in: the maximum temperature and if it is going to rain. It's all about personal comfort. We want to make sure we have the right clothes to keep ourselves at a comfortable temperature in the short dash between buildings and an umbrella to keep off the rain. That's about it. We live in a world cocooned in concrete, bitumen and metal. We live sheltered lives, seemingly safe from the inherent dangers of the natural world.

Closer contact with the vagaries of nature inherently brings a greater respect and awareness of its dangers. People who work on the land tend to be more aware of the impact of weather, through years of experience. Unable to encase themselves in cities and towns, they are at the mercy of the elements and have no choice but to respond and adapt to it.

But we need to pay much more attention to the weather if we are going to be able to assess bushfire risk. We are all a bit more conscious of water these days. If the garden is shrivelled and dry, if the dams are at record lows, if the local sports fields are brown and abandoned, then you can be pretty sure there isn't much water around anywhere—not in the ground, not in the grass and not in the trees. Everything is going to be dry and that much more combustible.

MEASURING THE RISK

Fire danger indices provide another measure of fire risk. The fire danger index is used by firefighters to decide how dangerous the conditions are for fires—how likely they are to get out of control once they have started. There are two fire danger indices in widespread use.

The forest fire danger is essentially calculated on the basis of how much moisture is in the air, how hot the air is, how much wind there is and how long it has been since it has rained. The forest fire danger index ranges from 0 to 100. Anything over 50 is extreme risk. Anything over 100 is off the scale and into uncharted territory.

The grassland fire danger index is calculated slightly differently. While it also takes into account the temperature, wind speed and humidity, it uses a more direct method to measure the amount of moisture in the fuel. While the forest danger index uses soil moisture as a surrogate for this, the grassland fire danger index includes a 'curing' factor or a measure of how dry the grass is. It also includes a measure of the amount of fuel per hectare, which is much easier to estimate accurately for a paddock of grass than it is for a forest. Basically, this measures how tall the grass has grown, on average, in the season.

Publicising the fire danger index for different areas might help those who are paying attention to the weather to get a clearer idea of just what the risk is.[2] The weather bureau produces excellent daily fire weather summaries for the fire agencies, but this information is difficult for members of the public to access.

However, no amount of extra information will help those who don't worry about the weather in the first place. If drought, high winds, extreme temperatures and fire bans don't ring alarm bells, will another scale, a better graphic or an extra adjective make that much difference? Is 'catastrophic' really that much worse than 'extreme'?

JUST ANOTHER FALSE ALARM

Leaving early, certainly before lunchtime, on every high fire-risk day may be inconvenient at times and require a bit of juggling, but I never found it particularly onerous. It was simply a matter of making a decision and sticking to it.

I did feel silly, though. Nothing ever happened. There was never a fire in my area and I could just as easily have stayed at home, like most other people in the area did. Nobody ever criticised my decision; indeed, people sometimes commented how good and sensible I was—in a solemn, slightly condescending way that implied that they would do the same thing if only their lives weren't so busy and full. No-one else I knew left on high fire-risk days. In fact, only around 3 per cent of people actually leave every total fire ban day. For all I was sure I was doing the right thing, I slunk home each time feeling slightly embarrassed for overreacting.

It is difficult to leave your home when you know there is very little chance of anything happening. Cyclones and floods cause more property damage in Australia than bushfires. Most houses are more likely to burn down from a faulty toaster or electric blanket than a bushfire. Even if you live in a high bushfire-risk area, within 50 metres of bushland, the risk of losing your home is extraordinarily low. On average, the equivalent of 83 houses in Australia are lost to bushfires every year out of 340 000 that lie in high-risk areas.[3] On the basis of those figures, the probability of your home burning down in a bushfire is just 1 in 4000.

With such small likelihood of disaster, it is easy to feel that it won't happen to you. Indeed, it probably won't. Then again, you probably won't die in a car accident (about 1 in 18 000) or die from skin cancer (about 1 in 29 000) but you still put your seatbelt on and use sunscreen. And it is even less likely that you will win the lottery (where the odds vary from 1 in 1 000 000 to 1 in 50 000 000) but thousands upon thousands of people buy tickets every week.

Humans are notoriously bad at estimating risk. The real reason we still buy lottery tickets, despite the extraordinarily low chance of winning, is because of our inbuilt optimism—we like to think that good things will happen to us and bad things will not. We wear seatbelts in cars and use sunscreen not because we are inherently sensible about reducing avoidable risks but because of concerted and persistent publicly funded campaigns to reduce the road toll and cancer incidence, with the added incentive of fines for non-compliance in the case of seatbelts. Changing people's behaviour takes a lot of time, effort and money. These campaigns have been successful because a concerted effort and investment was made to change people's behaviour over a long time.[4] They were also successful because, just like buying a lottery ticket, it costs us very little to put on our seatbelt or sunscreen.

The more difficult or expensive or inconvenient it is to change our behaviour, the harder it is to persuade people to do it. And preparing for a bushfire, which may never happen, is often difficult, expensive and inconvenient. Leaving early seems like an easy option, but if you have to rearrange child care, work or just daily schedules, it becomes more difficult.

Pets and livestock add an additional layer of complexity that is hard to resolve. It might be arguable that we should leave our homes to take their chances but not all pets can be taken out of harm's way and livestock can very rarely be moved to safety. We might be able to pack the dogs, cats and budgies in the car but what about the chickens, the horses, the goats and the sheep? Trading off our safety against theirs is much more difficult. We've seen farmers moved to tears by the loss of their cattle. For others, the loss of a pet becomes an underlying trauma that cannot even be spoken of within the family. Guilt and responsibility, love and care are powerful motivators in human behaviour and should not be discounted. Many deaths in bushfires occur when people are attempting to rescue or protect livestock.[5]

An even bigger problem with leaving early is that sometimes you can't. Fires don't always start after ten in the morning, a long way away or on a total fire ban day. One might start at night, on your property or on a day that isn't classified as high risk. The fire that burns your house down might well be the neighbour's burn-off gone out of control. Unlikely, but quite possible. Even when we plan to leave, we still have to prepare to stay—just in case. Or we may just want to give the house its best chance of survival, even if we aren't there to actively defend it.

PREPARING TO STAY

There are many good guides to preparing for fire. All the state fire agencies produce a wide range of booklets, DVDs and guides to assist the public to prepare for fires. To varying degrees, there is information for farms, tourists, homeowners and small businesses. It can usually be found in a variety of languages and with varying degrees of detail. Joan Webster's classic book on bushfire safety, inspired by the Ash Wednesday disaster, contains a great deal of helpful information for the householder. More recent books, like John Schauble's 'survival guide', provide clear, simple, straightforward advice.[6]

Preparing a property to survive a fire isn't necessarily high tech. A lot of preparation is incidental. People rake the leaves and sticks up from around their house just to keep it tidy and make it look nice. Gutters are cleared of leaves so that the water doesn't overflow or go to waste. Many residents install water tanks for additional household water. In the country, pumps may be put on dams to provide water for a garden or crops. Coincidentally, all of these habits promote fire safety.

Preparing your home

Like other fire authorities, the Country Fire Authority of Victoria provides many guidelines on how to improve the chances of your house and surrounding buildings surviving a bushfire. For an existing house, these include the following advice:

- Protect underfloor spaces with non-combustible sheeting or metal mesh. This will prevent embers from landing under the house and starting small spot fires.
- Cover all external vents with metal mesh to prevent embers from entering your home.
- Protect evaporative coolers with metal mesh screens. You will need to check with your evaporative cooler supplier to ensure the performance of the system is not compromised by installing the mesh.
- Ensure any external timber cladding is regularly maintained and all gaps are sealed.
- Place weather stripping around the inside of doors and windows to eliminate any gaps.
- Make non-combustible fire screens to cover external skylights.
- Fit leaf guards to gutters to prevent buildup of leaves.
- Install fencing made from non-combustible materials such as metal or brick.
- Ensure that fuel and chemicals are properly stored and not surrounded by fine fuel.[7]

Preparing a property to survive a fire isn't necessarily easy, either. Some activities are more time-consuming and costly. Every year, we have made more and more improvements to our property, trying to make it safer in the event of a fire. Our timber fascias are covered by metal cladding, our roof tiles are sealed underneath and any gaps filled with non-combustible insulation. There are sprinklers over every window and copper pipes that trickle water over the roof. We keep the area around the house clear of vegetation, wood or the other detritus that accumulates in everyday life. Our dam now has enough water for firefighting, with both an electric pump for everyday gardening and a petrol pump as a backup for when the power goes out. The pump house is concrete, with its own sprinkler to keep it cool. On every new project, on every new feature, the same question recurs: 'Will it burn?' Slowly over time, our house becomes more streamlined, more sealed and safer.

WATER

One of the biggest issues in house protection is availability of water. Some firefighters would boast that, provided they have water, they can defend any home. That may be tempting fate but it is certainly true that, without water, it's pretty hard to defend anything in a fire.

One of the earliest lessons to learn in house protection is not to depend on mains water. On the urban fringes, homes are often on the end of the line. With every house in the neighbourhood drawing on the supply, plus a couple of pumpers, it doesn't take long before the water in the tap dries to a trickle. Usually just when you need it most. Filling the bath, basins and buckets with water is a way of ensuring that water is there when you need it. You don't necessarily need a lot of it, although if you don't have much, it needs to be applied with precision. Many a house has been saved in a moderate bushfire with mops and buckets.

Swimming pools, rainwater tanks and dams are all good ways of ensuring that you have a supply of water on your property at all times. All you need is a good way of accessing and distributing the water. Pumps are popular but, like the mains water, you can't rely on the electricity supply either, so electric pumps have limited value in a fire. Much has been made of the failure of many pumps in the Black Saturday fires. Are diesel pumps better than petrol? Did the petrol evaporate in the heat? No everyday garden pump is designed to operate under extreme heat. Pumps need to be protected from radiant heat just like any other vital piece of equipment. Certainly we have to make sure there is no fuel around them to burn but maybe they also need to be housed in a solid, fireproof building—a small brick shed, for example. Pumps kept cool with a sprinkler on their shed roof seem to have performed better than those without.

There are few scenes more ubiquitous on the evening news in fire season than the sight of someone in a singlet and thongs, halfheartedly hosing down their roof as they watch the fires approach in the distance—the fiddler on the roof. Injuries from falls are all too common during fires. Getting up on your roof, exposed to both heat and wind, is not the safest option. What is the point? Most people's roofs are watertight from above. If water can't get in, neither can embers. Most people's roofs are made of non-flammable materials (unless you happen to have unusual wooden shingles). Provided the roof capping itself is sealed, there is not a great deal of risk from fire to the roof itself.

But, unlike water, fires move upwards. We would be better spend more time on the ground, aiming the water up, than on the roof, aiming the water down. Spaces under eaves are a classic place for embers to lodge or gain entry to the roof cavity. Making sure there is no access point upwards is far more important.

Many people also spend a great deal of money on sprinklers for their roofs. For tiled roofs, where there is always the risk of a tile blowing loose, wetting the roof may be a good thing. But you have to consider the wind. All too many sprinkler systems work fantastically on the calm summer's day when you first try them out. Under the gale-force conditions of a fire, they are just as likely to deposit all your water quite neatly on your neighbour's house, leaving your own high and dry. Sprinklers need to be carefully designed and targeted to ensure that they work the way they are intended.

On the ubiquitous corrugated iron roofs in Australia, there is even less need for sprinklers. Provided the roof is well sealed and in good condition, it is fairly unlikely that embers will penetrate. If you have loose sheets on your roof, you'll probably find it more cost effective to invest in some roofing nails than a sprinkler system. Of course, there is always the risk that the roof will blow off in its entirety but, in that case, the sprinkler isn't going to help. Never rely on just one system—always have a backup.

Sprinklers may be of more value targeting particularly vulnerable parts of the house, like the pump, or the windows. They are also good for keeping the garden damp around the house. They keep temperatures down, reducing combustibility, and wet embers that fall around the house. Many people opt for sprinkler systems around their verandahs, protecting a part of the house vulnerable to ember attack.

Windows often represent one of the weakest points of the house. The beautiful architectural picture window of a modern house may be an opening to a nightmare in a fire.

Windows, unlike solid walls, are at risk of shattering in the heat, or being broken by falling branches. Wooden frames may burn, aluminium frames may melt and steel frames may warp. Rubber seals can also catch fire, while plastic gives off toxic fumes and allows smoke to enter.

More important is the nature of the glass itself. Double glazing doubles the level of protection offered by windows. Even if the outer layer breaks, the inner layer may remain intact, denying entry to embers and flames. The thicker and stronger the glass, in general, the better it copes with heat, and there is a wide range of fire-retardant and fire-resistant glasses that perform very well at high temperatures, for a price. If you can't make the glass safer, then you need to protect it from heat, with water, shutters or even a sheet of sisalation.

The windows in my 1970s home are as thin and fragile as they are large. It would cost me a fortune to replace them with something safer. For the moment, the best I can do is to protect them with blinds and sprinklers and keep the area around them free from fuel.

TOXIC FUMES

A well-prepared and defended home should, in theory, at least in most fires, be able to protect you long enough for the most extreme fire conditions to have passed. Even if our homes catch fire, we all hope we can shelter inside long enough to be able to flee safely to a clear, burnt area of ground. In the past, fire authorities have thought that 10 or 15 minutes is long enough for a fire front to pass. Ten to 15 minutes inside, before the temperature outside drops enough for people to safely return outside and continue defending their homes from ember attack.

In the Black Saturday fires, many people had to shelter inside their homes for much longer. One resident of Steels Creek, whose well-prepared home survived extraordinary fire conditions, had to shelter inside for two hours before it was cool enough to go outside. He was fortunate. His aerated-brick home, with double-glazed windows and sprinklers, was able to withstand the heat and embers, even though it stood on top of a ridge and close to vegetation. Even so, the cars in the garage alongside caught fire, with no-one outside to defend them. The intense heat from the cars began to warp the door to the carport, allowing toxic smoke to enter the house. Even without fire inside the house, smoke can present a significant risk to residents sheltering inside—all the more so when the materials burning around a modern home are so toxic.

Compared to just a few decades ago, our homes are filled with synthetic materials. Not only do these burn more readily than many natural or more traditional materials, the fumes they produce are often highly toxic. Polyurethane and polyvinyl chloride (PVC) products release dioxins when burnt and are extraordinarily difficult to extinguish, adding to the problems created by smoke.

Sheltering in a house for 10 or 15 minutes is not likely to put anyone at risk from smoke inhalation. Ideally, the house will not have caught fire and the residents will be able to keep it that way when they return outside. Sheltering in a house for two hours is a different matter altogether. It means that the house has to be able to passively resist fire for longer, without active defence. And if the house does ignite, it may not have been able to protect us long enough for the danger from outside to have passed.

If the fires we face in the future are as severe as the ones on Black Saturday, we need safer houses and safer environments in order to withstand them.

GARDENING WITH FIRE

I love gardening and I love gardens, but gardening isn't particularly easy where I live. Locals call it 'gardening against the odds'. We live on the southern tip of the Kinglake Ranges, a rocky outcrop of sedimentary mudstone poking out of the rich alluvial soils of the Yarra Valley on one side and Plenty Valley on the other. The soil is shallow and poor, dominated by clay and shale. In wet winters, the whole district turns to slippery, yellow mud. In summer, the clay bakes hard and standard gardening tools must include a pickaxe and crowbar. Without copious and repeated applications of mulch each year, almost nothing would grow.

We have no town water, so the garden must compete for water with our own needs. Drinking water collected in the tanks makes its way through the house and into a grey water collection system before it goes onto the garden. A small dam is essential for additional summer water, particularly for the vegetables, but this water must also be saved for firefighting.

With no fences and surrounded by bush, our garden is a mecca for rabbits and wallabies. We've managed to bring the rabbits under control, largely by removing their warrens and the cover they need to shelter in. The abundant grey kangaroos rarely do much damage, other than keeping the grass down, although sometimes their romantic enthusiasms lead to the destruction of a young tree or two. The secretive swamp wallabies have a bigger impact. They have a particular passion for roses and apple trees; other than that, their targets seem quite random. A bush that has grown happily for months, even years, without attracting their attention, will suddenly be reduced to a pile of bare twigs. The only protection is in numbers and size. Mature, full beds seem to attract less attention than the more sparsely planted new beds on the perimeter of the garden.

I also love native plants and once dreamt of an entirely native, even indigenous, garden in the midst of our bush block. But in the end I'm too lazy and sentimental for that. I couldn't pull out the hardy survivors of the garden that was here before—the hollyhocks, roses and alstroemeria lilies that struggled on through grass and weeds alongside gnarled, old, neglected apple and plum trees. Even the much-despised, bush-invading agapanthus finds a carefully regulated place in our garden. The garden has ended up an eclectic cottage mix of anything that is tough, resilient and looks good—old-fashioned lavenders and salvias compete for space with modern agaves and echiums, alongside native croweas, correas and mint bushes.

Climate change has also changed Australian gardening habits. Once we treasured lush green lawns watered weekly, even daily, with the sprinkler and spent every morning spraying water with gay abandon over thirsty perennial borders. Now, we can only water every second Wednesday for two hours in the middle of the night. Drought-tolerant plants and mountains of water-saving mulch have become gardening staples. Our gardens are no longer lush, green and exotic. They are dry, hardy and locally appropriate. That is good for the environment. It is also good for fire.

By the time I look for plants that are low maintenance and drought tolerant, will cope with clay soils (waterlogged in winter and baked hard by summer), aren't weeds, don't get eaten by wallabies and rabbits—oh, and look good—the list of potentials has narrowed considerably. Add to that a preference for native plants and the list declines still further. If you want a fire-safe garden, the plant list shrivels to almost nothing. Indeed, the very concept of a drought-tolerant native garden seems to be almost anathema to fire safety.

The bushfire garden makeover

After the bushfires in Canberra, a garden makeover program advertised that it would be making over a garden burnt out by the fires. Fantastic, I thought, this would answer all my questions about gardening in a fire zone.

The selected house stood alone between burnt-out neighbours. Why its neighbours had burnt down and not it was a bit of a mystery. Just another one of those lucky breaks, a mini-miracle, so often left behind by fire. The owners hadn't had much time for gardening before—a bit of lawn, not much by way of trees or bushes. They were happy to have the makeover people there to pep the place up a bit.

By contrast, the unfortunate neighbours seem to have been a bit more keen on their gardens. The skeletal remains of trellises and shrubbery stood in silent testimony to their enthusiasm. I waited for someone to mention the role gardens can play in providing the fuel for fires, the risks of planting dry plants too close to the house, the way mulch provides a flammable carpet leading right up to the building. No-one did.

With typical gusto, the gardening team dug up new beds right under the eaves of the house and planted neat little flammable trees by the front door. The garden was mulched, shrubbed and vegetated to within an inch of its life with small-leaved, fine-fuelled, drought-tolerant, low-maintenance plants. It was green, it looked great—and it looked to me like it would almost guarantee that, next time there was a fire, this house would burn down too.

HOBART AND CANBERRA

Bushfires belong in the bush, not the city. Houses usually catch fire from the inside. They are just not particularly flammable on the outside. Flat, vertical walls and windows are not that easy to set alight—unlike curtains and blankets and stoves and sofas. To set fire to a house from the outside, you need extra fuel—some kindling or a bit of accelerant to get things going. The small, paved courtyard of an inner-city apartment is unlikely to provide much fuel for a fire, unless it is stacked high with newspapers for recycling. A pile of leaf litter swept up under the corner of an open slatted wooden deck is ideal; woodpiles stacked against the house are perfect. A large garden has the potential to provide much more fuel for a fire. Bring the bush into your garden and the bushfires will come to you.

When fire spread into the suburbs of Hobart in the late 1960s, most of the houses were on small blocks with neat, pocket-handkerchief green lawns, bordered by a narrow strip of exotic flowers and shrubs. Phil Cheney, a CSIRO scientist who studied this fire, noticed that the gardens provided relatively little fuel for the fire, which eventually—inevitably—extinguished itself as it drove closer into the non-flammable hard surface of the built-up town.

Thirty years later, when fires approached Canberra, the landscape was very different. The modern city of Canberra is beautifully designed with great swathes of native bush sweeping in between the suburbs, fingers of green that make the city a lovely place to live. Gardens have changed too. The surge of interest in native gardening that began in the 1970s has continued to increase the proportion of native and indigenous plants in our gardens, greatly enhancing their value as wildlife refuges and encouraging biodiversity.

Phil Cheney noticed the difference when he compared the fires in the suburbs of Canberra with those in Hobart. When the fires swept into the Canberra suburbs, they travelled much further than they did in Hobart or in other fires around townships. The fires' proximity to bushland and pine plantations seems to have been a factor; so too could have been the number of homes left undefended as their owners fled ahead of the fire. But just as important was the role of the gardens—providing a source of fuel for embers as they fell, kindling new spot fires that soon ignited fences and trellises, exposing buildings to direct flame contact and eventually setting them on fire. Familiar scenes of 'exploding' houses soon filled our television screens. This fire was carried by the gardens, not the forest—the forest threw the sparks in but it was the gardens that burnt down the houses.[8]

THE WELL-WATERED GARDEN

Yet, for every story of a garden catching fire and burning the house down, there is another story. Like the story of the couple in Humevale on Black Saturday who stood, hoses and mops at the ready, in black and swirling smoke, waiting in fearful trepidation for the first onslaught of embers and sparks. Their beloved garden, tended, weeded and watered over so many years, was to be sacrificed to the flames. This was no ordinary garden: neighbours described it as a botanic garden. Filled with deciduous trees, perennial borders and flower beds, the garden was their pride and joy. They knew they could not protect it from the flames; the best they could do was try to save their house.

The couple shielded their eyes against the howling winds, and heard the roar of the fire like the sound of an approaching jet aircraft. Suddenly, out of the darkness, flames spurted from the surrounding forest, leaping 20, 30 metres into the air, a giant, angry beast hurling its fury skyward. The hoses dropped from senseless fingers. There could be no defence against this. They turned to say their last goodbyes, clinging fearfully to one another. The fire roared forward and suddenly parted like the Red Sea, sweeping to the left and right and away. Everything was burnt and black in its wake.

But not this house. And not this garden.

There have been many examples like this, where neighbouring properties have been lost, where homes in the middle of cleared paddocks burnt down, and where even fire trucks, pumps and water have not been enough. And yet a relatively unprotected house in the middle of a lush garden has been spared. Were they just lucky to be on the edge of the fire? Was it mere coincidence, another one of those seemingly random events where one house is lost, while another is saved? Or can gardens play a role in protecting houses?

Time and again, well-watered and well-vegetated gardens appear to have protected homes from flames. Joan Webster, who wrote extensively on the Ash Wednesday fires of 1983, noted many houses in the Mount Macedon areas in particular whose spectacular, often deciduous, gardens were spared when all around them burnt.[9]

Fires are not always an unstoppable, uncontrollable force. Sometimes they can leap over roads and rivers, shower sparks and embers that start spot fires kilometres ahead, and burn down the most well-prepared homes. But even the fiercest fire will sometimes stop, seemingly on a whim, at a wire fence-line or a narrow ditch. Even the slightest change in vegetation can change the direction of a fire. Like water flowing down a hill, fire takes the path of least resistance. The leafy, green garden, next to the tinder-dry bush, offers just such a small resistance. Sometimes it is enough to part the flames, leaving many a manicured garden sitting untouched in a blackened landscape.

TO BURN OR NOT TO BURN

There is no such thing as a plant that doesn't burn. Get any plant matter hot enough and it will inevitably combust. The drier the plant, the quicker this combustion will be. For this reason, some fire authorities are loath to provide lists of fire-resistant plants—fire resistance is not just about the type of plant, it is also about where it is growing, its shape and location, and how well watered it is.

None the less, there are certain characteristics of plants that make them more or less likely to burn. For many Australian plants, characteristics that have adapted them to drought have also made them more combustible. Many of the features that reduce water evaporation promote combustion. Dry, tough, oily leaves rather than soft, moist leaves. Fine, feathery foliage rather than broad, shiny leaves.

Not all Australian plants are highly flammable. Even in this dry continent there are many plants that are not drought adapted—the plants of the rainforest, swamps, wetlands and alpine regions may all be adapted to wet conditions. With a higher water content, it takes longer for them to reach the necessary level of desiccation required for combustion. Plants with plenty of soft, leafy foliage are less easy to burn than those with an abundance of dry, woody matter. Many of these plants sport broad, shiny leaves, offering fewer edges for fire to catch. Even a sheet of newspaper is hard to set alight if you hold the match in the middle of the sheet—the edges catch much more readily. The more water a garden consumes, the less flammable it is likely to be. But, in this age of water restrictions and dry climate garden, a lush, semi-tropical garden is not always easy to achieve.

Not all drought-tolerant plants are highly flammable. Some species, particularly coastal plants, save water by having a high salt content. The native saltbushes and sun roses have grey-green leaves filled with moisture. Members of the pigface family or portulacas, whose cheery flowers brighten summer gardens and sand dunes alike, fill their leaves with salty water. The echiums from Africa and southern Europe also have this feature. Indeed, any plant that stores water in its leaves or stems is likely to take longer to burn than other, drier plants. For this reason, succulents are an obvious choice for a drought-tolerant, low-flammability garden. Many plants with grey-green foliage seem to share the characteristic of being both drought-tolerant and less flammable. Unlike their rainforest counterparts, they are able to survive with less water by storing it and using it efficiently but, just like any other plant, the drier and more desiccated they are, the more flammable they become. It is plant maintenance, rather than plant selection, that is crucial in a low-flammability garden.

Every state has its own list of plants that are high risk and those that are lower risk.[10] Many native shrubs, like wattles, bottlebrushes and banksias, are highly flammable. Other native plants, like the tough-leaved flaxes and lilies, are low fire risk. But their risk depends more on where they are grown and how they are maintained than on any inherent feature of the plant. If in doubt, cut a bit off and burn it. You'll very quickly find out which plants burn the fastest.

FIRE SAFETY BY DESIGN

There are many ways in which fires attack houses. We tend to assume that houses burn down through direct flame contact as the main fire front passes, but by far the most common way in which houses are burnt down is by ember attack.[11] Small particles of burning debris from the main fire fly metres, sometimes even kilometres in extreme conditions, ahead of the fire. If they land in a suitably flammable environment, they will smoulder, then kindle and roar into life. Ember attack can be brief or prolonged, depending on the nature of the fire. Fire survivors often describe 'red rain' as embers fall all around them, the wind forcing the deadly hail into every nook and cranny, into houses and roof spaces and certainly throughout the garden. Imagine wandering around the garden flicking lit matches at random into the beds. Surprisingly, not many of them will catch fire: most will simply burn themselves out. Campers all know how hard it can be, sometimes, to set a fire. But given the right conditions—a hot, windy day and a pile of dry grass—and even a spark can be enough. Spot fires caused by ember attack, not full-on fire fronts, are responsible for more house loss than any other type of fire. And spot fires commonly start in gardens.

Fortunately, controlling for ember attack and putting out spot fires is one of the easier tasks in firefighting. Attempting to control an approaching fire front is an entirely different story.

Notwithstanding the advance guard of flying embers, the main body of a fire is essentially the infantry—the ground troops. Unlike ember attack, a full fire front is generally not extinguishable. Even firefighters and trucks cannot tackle a fire head on. The head of the fire is where the fire is at its hottest. Most of us can tolerate the heat of a 1000-watt bar radiator fairly comfortably, even to the point when it singes clothing and reddens skin. Multiply that heat by a thousand and you will have some idea of the heat produced by the head of a bushfire, even a mild one. A dumpload of water from a firefighting helicopter might he momentarily effective in diverting a fire front of this kind, but this a limited option. The only effective way to deal with an approaching fire front is to starve it of fuel.

Planning for fire is like planning for rain but backwards. Like the rain, you cannot stop it entirely but you can divert it and channel it away from the house and other vulnerable places. Non-flammable surfaces like paving, bare earth, gravel and concrete obstruct fire. Water and water-laden materials, like dams, swimming pools, ponds and wet gardens, are also barriers.

Unlike water, fire tends to flow up rather than down. It burns faster and hotter uphill than down (although downhill fires can be fierce and frightening too). The heat from fires travels up. If you want to light a candle, you hold the wick above the flame. Whatever is above a flame will be exposed to more heat, and more likely to catch fire. In designing the spaces around a house, we need to think about the fire flow. Imagine setting fire to each shrub in the garden. What else would catch fire nearby—the fence, the overhanging tree, the window awning? Keeping the fire on the ground keeps it small and more manageable, less dangerous. Don't give it a ladder to get up into the canopy of trees or, worse, into the roof space of your house.

We often assume that gum trees are the greatest danger in a fire. Their highly flammable oily leaves, bark and dropping limbs can indeed pose great risks during a fire but it is rare for fires to get up into the canopy of trees. The majority of fires run along the ground beneath the trees. It takes a very large, very hot fire to sustain a canopy or crown fire 15 to 30 metres in the air. Such fires typically only occur when a large area of forest with heavy undergrowth burns under windy conditions and particularly up a steep slope. Where the forest thins, or the slope changes, or when the wind dies down, the fire falls back to the ground, often leaving the canopy of the trees scorched from heat, but unburnt. Reducing fuel under trees and removing fallen litter and lower branches makes it much harder for a crown fire to sustain itself.

In traditional suburban gardens, we often plant beds directly against the house, sometimes to soften the hardness of the building, to shade it or to maximise the central, clear space available in a small backyard. In fire-prone areas, this is not a good design principle. Quite apart from the fact that pouring water directly onto the foundations of your house is not a particularly good idea, growing plants underneath eaves and walls (or worse, climbing up the walls), is akin to stacking kindling against your home, particularly if you have overhanging eaves. Houses are much better off having a path or cleared area directly adjacent to them. Just a metre or two can make all the difference.

Fire travels along flammable gutters of shrubbery or mulch between the garden and the house. Breaking the flow of these paths of flammable material with hard surfaces or water will help to break the flow of the fire, perhaps just enough to divert it somewhere else. Isolated island beds of plants are safer than long, continuous rows. The elegant rows of trees leading up a winding driveway to a country home are all but inviting fire in by the front door. Plantings on the most vulnerable sides of the house—downhill and in the direction of prevailing winds—require particularly careful planning. Lawns, open spaces, well-watered vegetable gardens and fruit orchards may be better options here.

Gardening for fire is an extreme sport. You can opt for a

hard, dry architectural option. Or you can try the lush, green, well-watered option. One thing you can't do is take garden design for granted.

Landscaping for fire protection

The Country Fire Service of South Australia recommends that you consider the following points when planning a garden for fire protection:

- Keep vegetation near buildings low hazard (such as lawns).
- Plant trees and shrubs at least as far away from buildings as their mature height.
- Avoid a continuous canopy or line of vegetation from the bush to the house.
- Place fruit trees and vegetable gardens on the most hazardous side of the house.
- Remove overhanging and dead limbs from trees and shrubs.
- Prune trees and shrubs vertically to 2 metres.
- Keep plants well watered.
- Keep areas under trees cleared.[12]

PLANNING

Over the years, as our property became safer and my children grew older, my plan changed. I decided that I probably would be able to defend our house against a fire. Undefended houses are much more likely to burn down than defended houses and many burn down well after the fire front has passed through. In the immediate aftermath of the 1967 Hobart fires, one man drove from Snug down to Margate. Most of the houses were still standing. Five hours later, only 40 of Margate's 120 homes were still standing. The fire front didn't burn these houses down—embers did.[13]

I asked my neighbours about their fire plans. Most of them were women who either lived alone or whose partners were usually at work or volunteers in the fire brigade. We all realised that, in the event of a fire, it would probably be up to us. We needed to take responsibility for fire protection ourselves, rather than leaving it up to partners. We needed to think about what was feasible, practical and possible—and what was not.

I was lucky to have some neighbours who had put a lot of thought and effort into their fire preparation and I could learn from their experience. It was useful too, to share ideas with less-experienced neighbours, many of whom went through the same phases in their fire preparation that we had gone through. New residents were often obsessed with alternative escape routes, just as we had once been. Sometimes it can take a while before people accept that the fire brigade probably won't be able to help them on a bad day, that there may not be time for a warning, that they might well be on their own. People hear what they want to hear.

It helps to talk over plans, to consider 'What if?' Things that seemed quite reasonable suddenly shift into a different view when a neighbour asks 'But what if it is night time?' or

'What if the fire comes from a different direction?' Fires are complex. Rarely do we get a practice run to see if our plans and preparations work. What works for one fire may not work for another. We have to be prepared to adapt to changing scenarios, to know what the second-best options are and where the real dangers lie.

On your own, it is all too easy to think you are prepared, to shove any doubts or concerns into a far corner of your mind. There are always other more important, more immediate priorities to contend with. We have a great capacity for self-delusion. And self-delusion is much easier to maintain on our own.

The very thought of bushfires makes us anxious and anxiety is something to avoid. Sometimes it is easier just not to think about it. One lay witness at the Victorian Royal Commission commented that if you thought about bushfires all the time, you wouldn't live in the bush. But if you're not prepared to think about bushfires all the time, maybe you shouldn't live in the bush. Plenty of people who have lived in bushfire areas for many years finally decide they have had enough of the worry and move. Last year a friend of mine, who survived Ash Wednesday and has seen many fires come and go around her house, decided that enough was enough. After years of preparing to stay and defend, from now on she's leaving early. She'd rather spend hot days in town with the grandchildren.

Bushfires are so rare that it is easy to ignore the risks. They are what Max Bazerman calls a 'predictable surprise'.[14] We know bushfires happen but we think they won't happen to us. We live in a state of knowing, but not knowing, about fires—a state of denial. We blame others (Aboriginal people or European settlers or conservationists or local councils have increased the risk or severity of fires). We ignore information

(I didn't hear the weather warnings; I don't read the paper; my town wasn't mentioned). We deny responsibility (the fire was too severe; nothing would have made a difference). We doubt our ability to do anything about it (I couldn't have done anything; no-one else did anything, so neither did I). Or we project our worries on to something entirely unrelated (money, jobs or family problems).[15]

Meeting with other people makes it more difficult to maintain self-delusions. It also forces us to drag out our assumptions and preconceptions and subject them to the scrutiny of others. It allows us to have our plans and concerns validated and supported. We are not alone in preparing for fire. There are lots of people who take fires seriously. Most of us need social validation of our behaviour. It's a strong-minded individual who can continue to prepare for an alien invasion that no-one else in the world thinks will happen. Being in a group of like-minded individuals keeps us motivated, helps us to share ideas and reminds us that we are part of a community that can face these threats and survive, no matter what the outcomes.

There is always more to do. Our shed is not well protected and neither are the cars, which seem particularly prone to catching alight in bushfires. I worry about our large windows and wonder if I shouldn't invest in metal shutters. I haven't done anything about our smoke alarms which, in the event of a fire, would undoubtedly set off piercing wails just to add to the confusion and stress and are completely unreachable up in our cathedral ceiling. I still wonder if we shouldn't replace our tiled roof with corrugated iron, making it impenetrable to embers. Last summer I finally put together a bushfire kit of overalls, boots, gloves and mask and goggles. I never wear it. It is hot and scratchy and uncomfortable. I guess I'll put it on if I have to—if I have enough time.

We do check the pumps and run through the sprinklers every high-risk day. We have instructions for the pump stuck inside the wall of the pump house. I can remember where all the taps are and what to turn on and off in what order if I sit down and think about it. But how much will I remember or forget in the heat of the moment? I should practise turning them all on more often but it wastes water and makes the windows dirty. And it makes me feel stressed, so I don't.

Who can imagine what it would be like to be in a fire? Who can predict how they would respond? There are limits, I think, to anyone's imagination. There are definitely limits to what I want to imagine.

Sometimes I just don't want to think about fires.

IV · RESPONSE

14 · NEAR MISS

We knew Saturday was going to be bad. The messages had been coming through loud and clear all week. The hottest day since 1939. The driest January since 1983. Even more dangerous fire conditions than Ash Wednesday, or Black Friday. Unprecedented fire danger. Fire danger indexes were off the scale. The predictions just kept getting worse.

We checked the pumps and ran through the sprinklers. Mike headed up to the fire station early, making sure there was a full crew of volunteers standing by for the first fire call. I stayed home with the kids, raking away leaves, watering the plants and generally tidying up around the place. The UHF radio scanner crackled intermittently with the usual VicFire traffic—false alarms, house fires and traffic accidents. We all hoped against hope for another 'false alarm' day, when the firefighters would come home from their stations trying not to feel foolish for their rightful caution, when the people would scoff again at the fire authorities for crying wolf.

It was not to be. The fires started fierce and fast as the day rapidly heated up and the north-westerly blew its hot wind across the land. There was a fire near Kilmore, directly to the north-west of us. With any luck the change would arrive before the fire, but it was still heading our way. The air conditioner struggled to keep the lounge-room cool. The kids brought the guinea pigs and the dog inside and I hosed the chooks down in between clearing any remaining flammables away from the house, keeping an ear out for more local activity on the scanner. The phone tree was activated—everyone needed to get ready.

The girls put on their jeans and long-sleeved cotton tops, grumbling in the heat. My youngest had filled every available water bottle in the house, offering them to her sister and myself on a regular basis. But the only one who would drink was the dog, who obligingly consumed three bowls of water and promptly threw up in her basket.

By mid-afternoon I heard the distant wail of the Panton Hill fire station siren and knew that they had been called out. The fires were much closer but not an immediate threat. They were kilometres away. We still had an hour or two, surely.

The first mayday call was made by a trapped tanker in a town quite close by. Their frantic calls for help were followed a few moments later by equally frantic calls to keep away. It was too dangerous to help them. We listened until we heard they were safe. It was the first of many mayday calls that day.

Suddenly calls came in for fires in Strathewen, Kinglake and Christmas Hills. We were surrounded. The sky went eerily orange. Fires were appearing much further ahead of the front than I'd expected. Our quiet road suddenly filled with speeding cars, loaded with mattresses and possessions. The phone rang constantly as neighbours saw smoke and had ash and embers landing in their gardens. The girls stayed glued to the computer, flicking between websites to glean more information. The cold front had arrived in the western coastal towns—Portland, Warrnambool, then Geelong. We held our breath and waited to see which would hit us first, the southerly or the fire. The cool change would inevitably spell disaster for some and salvation for others. Which ones we would be was a matter of luck.

A blast of ocean air roared through the trees, stripping their leaves and branches to skeletons, bringing momentary relief. A neighbour rang to warn me there was a fire on the road directly to my south, below our thickly vegetated gully. I fear a fire from the north, but I dread a fire from our south. Rubbish, I snapped at my neighbour, smelling for smoke, ringing another neighbour across the gully, directly to windward. There was no fire there—the danger for us was over. But I had no idea where Mike was when the front came through. The shadow of the lost Panton Hill tanker on Ash Wednesday lingered in the back of my mind. I stopped listening to the scanner.

As darkness fell, the phone rang incessantly. The fires lit up the northern hills like beacons. Many only became aware of the danger after it had passed. Someone had heard there were people dead in Kinglake. There had been nothing on the scanner about casualties. It was probably nothing. The power finally went off. The last thing I wanted in my house was the flame from a candle.

Mike finally came home, dirty, exhausted and overwrought. It had been awful; there were bodies and cars everywhere. There was nothing they could have done. It was worse than we could possibly have imagined. Relief at our good fortune was interspersed with a sense of guilt and distress for what had happened to others so close by. We fell into an anxious, fitful sleep, interrupted by phones and pagers as the cool breeze swept the heat and smoke, but not our fears, away.

In the morning we awoke to the news that while Victoria had burnt, southern Queensland had been flooded by its second week of torrential rain.

15 · FACING THE FLAMES

THE PSYCHOLOGY OF FIRE

The psychology of fire comes down to the psychology of survival. How do we respond to a life-threatening situation? Everyone is different; every situation is different. It is probably something that is impossible to predict unless you happen to have some extraordinarily high-risk occupation that puts you in life-or-death situations on a regular basis.

Imagine you are walking around the side of your house into your backyard. Your mind is filled with everyday banalities: what to have for dinner, who won the footy, how annoying your boss is. Your expectations, as you round the corner, are pretty average—a clothesline, a garden bed, maybe a kid or two. Instead, you are confronted with a lion. A rather large, smelly, annoyed-looking lion.

What is your first reaction?

Most people would probably stop dead in their tracks and just stare. If you were in a cartoon, you would have bulging eyes and exclamation marks popping out of your head. In effect, you freeze.

In this instant, the decision-making part of your brain is desperately searching for information on what to do. The clever, intelligent, thinking part of your brain is rummaging hurriedly through every thought, idea and memory labelled 'lion', to see if somewhere, in all your years of experience, there is a plan of action for what to do when a lion unexpectedly turns up in your backyard.

Nine times out of ten, it won't find one. Our brains try to fit all the expected things onto the incongruous image in front of us. An over-sized dog from the neighbours? A really big kid's toy? An elaborate hoax by a brother-in-law? Some stupid television show's idea of good ratings? Increasingly desperate, the brain often grabs whatever convenient plan of action happens to be lying around, something you usually do when you walk into your backyard. Maybe it is picking up a kid's discarded toy or putting the beer can down on the barbecue. Here, says the thinking part of your brain, do this—it's always worked before.

The behaviour of survivors from the New York World Trade Center collapse illustrates just these stunned responses. Despite the massive explosion that destroyed four floors and propelled the building in a sideways lurch, few people ran instantly for the stairs. A great many looked around to see how others reacted. Many rang family and friends for information on what had happened. Finally, once they did decide to leave, they turned off their computers, collected their handbags, and leisurely left the building.[1] The thinking brain can't compute what is going on, can't find a pre-existing pattern to match this situation and can't provide an appropriate response. It either ignores what is going on, denies that there is a problem or comes up with a feasible and normal explanation for what are clearly abnormal events.

In more immediately life-threatening situations, there is another part of your brain that swings into action. This is the older part of your brain, the bit that lies underneath the decision-making, thinking part. Some people call it the primitive brain. Generally speaking, it looks after our more basic motor skills and leaves most of the conscious decision-making tasks to the thinking brain. Until there is an emergency. If a bus comes speeding towards you down the street, it is your primitive brain that causes you to leap out of the way before you've barely registered what has happened. When the lion looks at you appraisingly and yawns as if he's giving his jaws a bit of a warm-up, the primitive brain slams the door shut on the procrastinating thinker upstairs, fires a message direct to your legs and shoots you down the lane and out the front gate, leaving Usain Bolt and the lion for dust.

When we are threatened, our heart rates lift, pumping richly oxygenated blood away from secondary functions, like digestion, and towards the vital organs of the lungs, heart and muscles. The blood thickens, ready to staunch wounds. White blood cells rush to the surface of the skin, ready to combat infection. Nutrients are released to aid the muscles. Our vision focuses down into tunnel vision, excluding all peripheral information. It is common for hearing to drop out as we focus all our attention on our strongest sense—vision. We are ready for action: to flee, fight or defend ourselves.

These physiological responses have saved countless lives through evolutionary history. For the last 2 million years or so, the hominid emergency response system has kicked in to save us from leopards and lions, charging elephants and club-wielding cousins. Many of these physiological responses are even older still, having descended through generations of mammals, back to shared ancestors with birds, reptiles and even amphibians.

These ancient survival skills have unexpected consequences in the modern world, particularly for an animal so uniquely dependent on its large and complex cortex—the executive control centre of the brain. In an emergency, this thinking part of the brain simply gets bypassed.[2] This means that the parts of the brain that decide what to lay down as memories, and what information to forget, aren't working. This is why our memories of disasters are often fragmented, a series of unconnected images or vivid, flash-bulb memories. In addition, the part of the brain we usually use for everyday problem solving is cut out of the loop. We find it very hard to think through problems logically, because our bodies are completely focused on survival.

If the threat is a straightforward one, like a lion, this may not matter. If the threat requires us to make complex decisions, we are in trouble. If we don't know what to do to survive beforehand, then we won't necessarily be capable of working out what to do in a disaster. Our decisions, while they may seem sensible, may be less than optimal. Survival skills in a modern world often have to be learnt; they can't just be worked out in the heat of the moment.

This is why we need emergency drills. We all think we will respond logically and sensibly in a disaster but we don't. Studies of people in all sorts of disasters, from aircraft accidents to building fires to bomb threats, have shown that nine times out of ten we sit like stunned mullets, denying that there is a problem, while warning bells ring, smoke fills the building and walls come tumbling down.

In one of America's worst building fires, the Cocoanut Grove nightclub fire in Boston, people didn't respond quickly to the danger. They watched in amusement as waiters attempted to put out the flames spreading across the ceiling.[3] They ignored calls from a bellboy to leave the building. They continued with their drinks and their dinners and conversations. Finally, when the danger was too much to ignore, it was simply too late. Many of the 492 people who died could not escape through the revolving front door, the blocked fire escapes and the boarded windows. A great many more died still sitting at their tables or while making phone calls.

'WE'LL BE FINE...'

This is a common feature of disaster responses. We imagine people will panic and rush to the door, like they do in the movies, but they don't. They respond with what is called the 'normalcy bias'.[4] They normalise the situation: they explain it away, they check with others, they stay sitting at their desks or at the bar.

When a huge, angry-looking smoke plume hung over Marysville in the late afternoon of 7 February, normalising was exactly how people reacted. They noticed the smoke but dismissed it as smoke from Kilmore; they talked about it with others who didn't seem worried; they got on with their everyday lives. Their rational brains saw no reason to be alarmed: it could all be logically explained away as nothing to worry about. There was nothing as yet immediately life threatening to disturb the underlying primitive parts of their brains. A few people were worried; some people did try to warn others—but they were the exceptions. It takes a certain critical mass for people to respond to a crisis. A few people could be wrong and we ignore them; a few more give you pause for thought; if everyone runs, we run too.

It is not that we are stupid, ignorant or plain perverse. Our brains are simply not equipped to deal with disasters. In order to know how to respond in a disaster, we need information, we need training and we need practice.

There are people who don't respond to disasters by normalising them. They know what to do. They assess the situation and leap into action immediately. They recognise the warnings signs for what they are. They know that a lion is a dangerous animal or that a visibly growing smoke plume means the fire is heading this way—and fast. They are often responsible for shaking everyone out of their lethargy. They take command, issue clear, loud instructions and make sure people follow them. They are often responsible for saving countless lives.

More often than not, these people have had some kind of training for emergency responses. They might be fire wardens who take their duties seriously. They might have military training. They might be firefighters or air hostesses. They have prepared for this eventuality, whatever it might be. They have thought through the options and know what the best plan of action is.[5] When disaster strikes, their brains flick through all those pages of behaviour patterns in their heads, find the one labelled 'lion' and read 'move calmly and slowly away, preferably out of line of sight'. They have a plan and they enact it automatically, without the need to think, consider, procrastinate or consult.

We may not be able to train for all emergencies that might befall us but we can train for the risks we know about. We can read the safety card in the plane and identify our exit points. We can familiarise ourselves with the fire plans in the buildings we use. Providing ourselves with the skills we need in a bushfire requires a bit more effort and consideration but it is not overly difficult. The more we put into it, the better we are likely to be able to cope, should we ever have to face the flames ourselves.

FIRE AND FEAR

Humans have a complex relationship with fire. We fear it and yet we are fascinated by it. Fire has long been an essential domestic servant and yet, with a careless moment, it burns down our houses and takes our lives. We are compulsive pyromaniacs and yet there are few things more terrifying that the sensation of an approaching bushfire.

Understanding what to expect from a fire is crucial to controlling our reaction to it. Fear is what kicks off our adrenal system, puts our primitive brain in charge of our survival and puts us at risk of making snap decisions that may not be the best or most considered options.

Fear is a part of our instinctive survival system. It is part of our ancestral heritage, so much so that some believe that different fears reflect our evolutionary history, as much as fossils in the rocks reveal the different ages of the Earth. We share our fear of loud, sudden noises with nearly all other animals. We share our fear of heights with all other mammals (and indeed, with many young bird chicks as well), placing its origin at least 140 million years ago. We share our fear of snakes with all other apes and monkeys, inherited from a common ancestor 20 million years ago. Our fear of the dark is also shared with other light-loving primates.[6]

More recent fears are species specific—independent adaptations to the particular risks their environment presents to them. A fear of open spaces, for example, is distinctive to humans but not necessarily other apes, perhaps because of a heritage of life on the African savannahs. Our complex brains provide another unique overlay to our fears. We learn fears; we acquire them. Sometimes these are useful fears, sometimes they are irrational—fears that have no adaptive value or that take our rational instinctive fears to an irrational and maladaptive level. An excessive fear of spiders, a fear of social contact, a fear of flying are all fears which in some individuals have become amplified and extended to the point where they interfere with normal life.[7] Learnt fears can also be unlearnt. We may not be able to stop ourselves from jumping in response to a sudden, loud noise but we can learn to control our fears. Our primitive brain will always respond to fear stimuli, but our thinking brain is able to deal with this response rationally, provided it has the tools to do so.

Fire in itself does not elicit fear in humans. Almost 2 million years in close association with fire probably makes a fear of fire maladaptive. But then, few other animals show a fear response to fire either. Some are actively attracted to small fires, recognising the opportunity for an easy meal. In other animals, curiosity is only curtailed by a burnt nose or paw. Small fires, at least, are not instinctively fearful.

Large fires are a different matter altogether. Large fires tap into some of our most fundamental fears. These basic fears trigger the survival mechanisms of the primitive brain, prompting us to freeze, fight or flee. None of these responses is useful in a bushfire. Knowing what will happen in a fire and knowing what will frighten us are critical for controlling our fear and responding rationally rather than instinctively to the flames. Knowing that it may be loud, dark, hot, difficult to breathe and filled with unexpected explosions goes a long way towards controlling our responses, keeping ourselves in charge and safe.

Feeling the heat

The human body has a remarkably low tolerance for heat. Exposure to heat is measured in kilowatts per square metre (kW/m^2). The heat from the sun on a hot day reaches a maximum of 1.0 kW/m^2. At 1.6 kW/m^2 we feel discomfort and tend to move away from the heat source. A little hotter and the outermost layer of skin scalds or blisters; fleeting contact with an iron or oven will cause this type of burn. At 5 kW/m^2 we start to feel pain and the first layer of skin is destroyed and the second layer is damaged. At 12.5 kW/m^2 wood ignites and an unprotected human will die.

Fire fatalities are rarely caused by direct flame contact. In house fires, most deaths occur from smoke inhalation. In bushfires, people can die from the consequences of the extreme heat radiating from the fire long before the flames reach them. Dehydration and overexertion, heart attacks and smoke inhalation all contribute to fire deaths as well as direct burns.[8] The heat radiating from a fire front can exceed 100 kW/m^2, multiplied across a fire front that might be kilometres across and, in extreme conditions, tens of metres high. The human body typically requires an internal temperature of between 40°C and 42°C. Once our body temperature goes above 42°C, our organs start to fail; our heart, lungs, brain, liver and kidneys are all prone to malfunction and severe injury, and death soon follows.

Our ability to withstand or succumb to heat depends upon a variety of factors. We can stand high temperatures for short periods of time, but not for long periods.[9] It all depends upon how effectively we can keep our internal body temperature down. But keeping our body temperature down can come at a cost. Humans use evaporative cooling—sweat—to cool themselves. If we don't replace the water lost, we run as much risk of death by dehydration as hyperthermia.

Damage from heat also depends upon how well heat is conducted; air does not conduct heat so well as water or metal. Bare skin feels the heat much more quickly than covered skin. You can measure this effect quite simply in front of a fire. Expose your bare skin and you can only withstand the heat for a short period of time. Clothes offer a surprising amount of protection—they reduce thermal radiation about threefold (assuming, of course, that your clothes themselves are made of natural, less flammable fibres).

Even more protection is offered by a building—the thicker and less flammable, the better. Which leads to the basic principle of fire protection—if the fire is inside, get out and, if the fire is outside, get in, at least while the fire front itself is passing.

LIKE A HUNDRED JET AIRPLANES TAKING OFF OVERHEAD

Everyone mentions the noise of a big fire—trains, trucks, motorbikes and jets—but all fail to do justice to the deafening sound of a fire approaching. Fires roar—'the roar was so great we had to scream to be heard' recalled the manager of the mill near Murrindindi in 1939.[10] Even small fires roar when you add light fuel or oxygen. Magnify that sound a hundred thousand times to imagine the sound of a bushfire approaching.

We can cope with loud noises. We voluntarily submit ourselves to them at rock concerts and workplaces. After a while, you don't even register regular loud noise. It is not the sound that is frightening. It is the unexpectedness of it. It is the approach of it, the build-up, the impending threat of something large and dangerous coming towards us. We use the same stimulus in movies to scare ourselves silly. A building crescendo of sound makes our hearts race, our muscles tense and we cling to our seats ready to leap out of them when the big, bad or awful thing finally arrives. Not knowing what the big, bad or awful thing is going to be is the really scary bit.

We all fear the dark to some extent. Humans simply don't

have good night vision. Vision is our dominant sense. Without it, we feel vulnerable and exposed. Darkness heightens all our fear responses, makes us jump at sudden noises.[11] Fires create the darkness of night in the middle of a sunny summer's day. They fill the air with smoke, restricting our visibility even when there is light to see and making it hard to breathe. Our familiar daylit world changes to a dark and unfamiliar place. We can't find our way around; we can't use our regular landmarks to navigate; we can't judge distances. In the dark and smoke we are isolated from our friends and family, alone in a strange and scary place.

Alone, in a dark place, with the roar of a fearsome monster approaching, we are primed for survival mode. Our primitive brain is just waiting to leap out and take charge. It just needs a trigger, one thing to set it off. Pop a paper bag behind a person in this state and they'll leap out of their skin.

In one Victorian fire, a well-prepared resident was defending her home against ember attack. She knew what she was doing and had her house defence under control. As she was putting out small spot fires around her home, a hay shed caught fire. Unnerved by this, the crash of a nearby falling tree proved too much and, despite the fact that neither event particularly put her in danger, she leapt into her car and drove through the flames to escape.

On reflection, the resident realised that by fleeing she had potentially put herself in much more danger than if she had stayed at her home, which did not burn down. But in her heightened state of alarm and anxiety, it did not take much to set off her instinctive survival mechanism. Being aware that we are in this state and that things do explode, crash and erupt in a fire go a long way towards enabling us to control our responses.

Knowing what the monster is removes much of the suspense. Our thinking brain works hard to rationalise our fears, to control them. If it knows what the noises are, what to expect, it will be able to control our emotional responses, lessening the risk of making dangerous choices. Knowing how to deal with the monster, that the monster can be dealt with, takes much of the fear away. We may still have adrenaline coursing through our systems and be ready to fight to save our lives but we are no longer at the mercy of our primitive emotions. Our rational brain has a chance to stay in control.

There are many unexpected aspects to fires. It is the very diversity and complexity of fires that makes them so hard to plan for. Many people underestimate the strength of the wind in a fire. Fires are most dangerous on windy days and large fires can create extreme localised wind events, mini-hurricanes, seemingly unrelated to the prevailing weather conditions. Some fires are preceded by a prolonged ember attack; in others, flames arrive suddenly, without warning. Many people expect the fire to arrive from a certain direction, like an orderly military attack, and are taken aback when the flames suddenly appear out of nowhere behind them, or when fire seems to erupt simultaneously all around.

You can prepare for the most likely type of fire to occur in your area but ultimately there is no way to predict precisely what form that fire will take under the given conditions of the day. You have to be prepared for anything, ready to cope no matter what comes your way. And there is almost no way of knowing how you as an individual will respond on the day—except for the fact that the more you prepare, the better off you are likely to be.

People who are prepared, who have a plan, who know what to do, seem to suffer from less fear and indecision in the face of a fire. They commonly report being too busy to be scared; they 'just get on with things' and 'put it all into action'. No matter what the fire brings, they remain in control of how they respond.

MEN AND WOMEN

Men and women generally respond differently to fires, indeed to any disaster. Women often report being more anxious about fires, while men report being more organised in planning for dealing with fires. Women tend to rate the risk of fire to their homes more highly than men rate it.[12] This difference could be due to women being more emotionally responsive than men. Alternatively, it could be due to less training or experience with fire among women. Among trained firefighters, males and females report similar levels of anxiety about similar parts of their jobs. But female firefighters report more anxiety about their skills than their male counterparts.[13] In a survey of south-western US communities, researchers found women to be more concerned but less knowledgeable about wildfires than men. Knowledge is power; it seems to put you in, or at least makes you think you are in, control.

The response of women to fire risk may also be compounded by the presence of children in a household. Children change how people perceive risk. People with children living near Love Canal in New York were much more likely to be aware of and worried about the risks associated with a nearby hazardous waste site than those who did not have children.[14] Any parent of teenagers will be all too aware that risks they once regarded as low to themselves are suddenly multiplied in magnitude when applied to their children.

This gender difference is probably grounded in our evolutionary history. By and large, human males are hunters and fighters. It is adaptive for them to think they are invincible and immortal. Their role is to face the danger. And if they fail, there are always other males to take their place.

In many social animals, females play a very different role. Biologically, females are simply less expendable than males.

Child-rearing is a time-consuming and labour-intensive activity. The more females present, the better the chance of survival and the number of offspring successfully raised. As a result, women tend to be more risk averse. They are better at recognising danger and less willing to face it. It is nearly always women who choose to leave early and men who decide to stay and fight the fire. Very often households choose to send a mother or grandmother and children away to safety while the strongest and healthiest men (and women) stay to fight the fire.

In other situations, the conflict is not resolved. Women who want to leave stay to help their husbands despite their better judgement. Some women make the difficult decision to leave, even though their husbands won't. If decisions are made without fully considering the preferences of different family members, plans come unstuck. One 'household' may have decided to stay but, when left alone, some members (often women) may decide to leave, perhaps at the last minute. In other cases a family has decided to leave but, when faced with fire, a husband insists on staying.

Such gender patterns are sweeping generalisations and there are, of course, many exceptions. But the pattern holds broadly true overall. 'Next time, I'll listen to my wife' is a recurrent refrain after big fires.

Unresolved differences in approach not only place plans, and lives, in jeopardy, but they also carry the burden of regret and recrimination. One husband who had planned to defend his home was persuaded to leave by his wife. Aerial television footage revealed their home, safe and intact, immediately after the fire had passed. By the time they were able to return, their home had burnt down, a victim of a smouldering ember that might easily have been extinguished if they had been at home.

Understanding the fundamental differences between individuals in their approach to fire is vital for effective fire planning. We have to recognise that we may respond differently, that we may have different expectations and the way we think we will respond to a fire is not necessarily the way our partners will respond. These differences are deeply rooted. They are often not a matter of choice. We have to understand, respect and work with these differences if we are to ensure that our carefully considered plans don't go astray.

Very often, household protection plans are determined by the male of the household. But in the event of a fire, the male of the household may not be home. In many farming communities, the local fire brigade may comprise most of the adult male population. It is all too common for firefighters to lose their own homes while defending someone else's. On a large property, farmers may be protecting stock or assets away from the house or on a neighbour's land. Very commonly, home protection rests with women.

During the Wangary fires on the Eyre Peninsula, Tish Proude faced a fire on her farm with little contact from her husband Robert, who was fighting the fire elsewhere. Whatever Robert's plan might have been for protecting their property, she did not know it. She told ABC journalist Ian Mannix that her plan had been to leave the farm if things got too dangerous.

> That was the plan in my head ... I didn't talk to Robert much about fires, but I knew he had a plan. There's a division of responsibilities on a farm, women do some things, men do others. I was worried about what would happen if Robert wasn't here.[15]

Her immediate response to the fire was to get in the car with her baby daughter and flee. But she knew how dangerous that was and stayed, successfully surviving the fire with help from her mother, brother-in-law and a friend.

Even on the urban fringes, this pattern holds. Mothers with young children are more likely to be home during the day than any other demographic group (except the retired and elderly). In the Ash Wednesday fires, many husbands were at work in the city. When the fires struck, their wives were home alone with the children, and their husbands had to negotiate roadblocks and dangers to get home to help them. While women may be more at risk driving out of the fire zone, men may well be more at risk driving into it, to protect their families or their homes.[16] During one fire in the Otways, an older gentleman bypassed a roadblock by walking along the beach to get back to his fire-threatened house, saving not only his own but also several neighbours' houses. In Strathewen in 2009, at least one man is thought to have died attempting to drive back to his partner trapped alone in their home. Yet despite the fact that women are the ones most likely to face a fire threat at home, it still tends to be men who make the decisions about fire planning.

All of these factors play a role in understanding how and why people die in bushfires. People die fighting fires, they die fleeing from fires and they die sheltering from fires. By far the greatest danger in fire is being caught in the open, being exposed to the radiant heat that we have so little tolerance for. Both firefighting and fleeing from fires places us out in the open, at greatest risk from radiant heat. But we cannot simply hide from fires either. Keeping safe from fires is a complex and difficult course of action. Fire may be rare, may be unlikely but, if it happens, we need all the skills and training we can get to maximise our chances of keeping ourselves safe.

FIREFIGHTING

Fighting fires is undoubtedly a dangerous activity. According to the National Interagency Fire Center in the USA, almost half of all the US wildfire fighters who have died between 1910 and 2007 (nearly 1000 in all) have died from burn-overs or from other flame or heat contact. A quarter died in vehicle accidents (including aircraft and helicopters), while heart failures, the next largest killer, account for just under one-eighth of all deaths.

Of the 689 people who died in bushfires in Australia between 1901 and 2007, 40 per cent were actively fighting fires. About half of those killed (123) were firefighters, while the remainder were general members of the community either actively defending their homes or assisting with firefighting.[17]

Bushfire fatalities are generally divided into two groups: firefighters and civilians. The statistics are compiled separately and analysed differently. There is an assumption that these two groups are fundamentally different—that firefighters are trained professionals and civilians are an amorphous mass of unknowns.

The reason for the distinction is largely bureaucratic. Firefighter deaths are an occupational health issue; civilians deaths are a population health issue. In the cities, with a dominance of paid firefighters, this distinction probably makes sense. In the country, where the vast majority of firefighters are volunteers and community members, the distinction makes very little sense at all.

The difference between a firefighter and a community member comes down to which hat they are wearing at the time. If they are wearing a uniform and on a truck at the time of death, they are a firefighter. If they were at home defending their own or a neighbour's property, they may be classified as a civilian. The seven men who died in the 1919 fires at Saltern Creek are classified as civilian deaths. If similar events had occurred today it is highly likely that many of them would have been volunteer firefighters.

With improvements in firefighter safety and training, we would expect the number of avoidable deaths to decline. But improved safety equipment doesn't always make for safer behaviour. Indeed, some driving specialists argue that the best way to make people drive more safely would be to put a large spike in the middle of the steering wheel. By making cars safer, and accidents more survivable, we may well have simply provided an opportunity for some drivers to be more reckless and dangerous.

Does this apply to firefighters? Research by Phil Cheney into American fire fatalities suggests that the deployment of fire shelters may have encouraged managers to place firefighters into riskier situations.[18] In the recent Kilmore fire, one fire tanker crew justified proceeding up a dangerous mountain road despite being told the risks, on the grounds that they had a safer dual cab tanker to shelter within. As events transpired, they were lucky to survive. Would they have made the same decision if half their crew had been sitting in the open back of an older-style tanker? The trouble with safety improvements is that we automatically take them into account when calculating the level of risk we are prepared to accept. If trained firefighters have trouble calculating risk, how much harder is it for a member of the general public who may have little or no knowledge or experience of fire?

16 · LEAVING LATE

When is a threat not threatening? When it hasn't happened yet.

As we've already seen, people tend to only recognise a 'threat to life' when the evidence is smacking them in the face—a clear and present danger. In the case of fires, the event that people recognise as a 'threat' is not a warning, a total fire ban day, smoke or even embers. People typically don't feel threatened until they see the flames.

If you don't intend staying to defend your property during a fire, the official advice is to leave early, the night before or in the early morning of a high fire-risk day. The vast majority of fires in Australia start in the mid-afternoon, so leaving in the morning is generally a pretty safe option.

But leaving your home because of a bushfire risk, when there is no fire within cooee, does seem excessive—an over-reaction—so, instead of choosing between 'stay or go', most people adopt a 'stay and see how it goes' approach. They intend to leave, but only once there is a fire in the area, or once a fire threatens them, or if the fire gets too much to handle.

Maybe this wouldn't be so bad if they were actually paying attention to what was going on around them. If, on a high fire-risk day, people were listening out for fires in their area, checking their scanners, getting the car packed and animals organised, they probably would be able to leave safely at the first sign of danger. But they aren't really prepared to leave at all. What happens is they stay (ignoring, denying or rationalising the threats away), until the threat to their own lives overwhelms them and then they flee.

Part of the difficulty lies in the dilemma between saving your house and saving your life. Undoubtedly the best way to save your life is to leave the bushfire area on high fire-risk days. Don't be there. The best way to save your house in the event of a fire is to be there to defend it. Houses are much more likely to survive when they are defended, and highly likely to burn down when they are not.[1] Obviously we want to save both our homes and our lives but the emphasis we put on one over the other depends upon how dangerous the fire is. With a low to moderately severe fire, the risk to life is low, while the risk to house is high. People are more likely, under these conditions, to stay and defend. As the fire severity increases (as it may well appear to do during the approach of a fire), the risk to life increases. At some point, people switch strategies, from a house-saving option to the lifesaving option, even though leaving at this point may actually place them in grave danger. All too often people drive through flames and smoke to leave their house, which survives the fire unharmed. All too often those who fled were not so fortunate. It is sadly all too common for people to flee their homes into danger, sometimes to their deaths, only for their house to survive intact.

We know that the greatest risk in a fire is being caught outside. Before 2009, over 70 per cent of all fire deaths occurred outside, in cars or in the open.[2] Only around 13 per cent of bushfire fatalities occur inside buildings. Even after Black Saturday, with its unprecedented death toll among people sheltering in houses, the proportion of fatalities inside buildings may rise to 25 per cent, but this is still far fewer than those caught in cars or in the open.

The thicker, more insulated and less flammable the material we can shelter behind, the better. Leaving late, when fires are approaching, takes us away from the relative safety of a building, and into the open. Here we are simply relying on good luck and good fortune to protect us. Many people survive fleeing late; just as many people survive staying put. Of those community members who have died in bushfires in the past, around one-third die while leaving late.[3]

ON THE ROAD

It is not just people who live in the bush who are at risk from bushfires. People come to the bush for other reasons: to visit, to work or just to pass through on the way to somewhere else. In summer, millions of people travel through our national and state parks. These areas are popular tourist destinations. Many people drive in cars. Others prefer to trek along isolated walking tracks to appreciate the quiet beauty of the Australian bush. Getting away from it all, being out of touch with the outside world for a while, is part of the attraction of these walks.

Some parks, particularly the very popular ones in Tasmania for example, keep track of visitors walking the longer trails and restrict the numbers of trekkers at any one time. Most other parks and trails are open to anyone who wants to visit and their popularity sometimes takes park managers by surprise. On any one day an unknown number of people are walking through bush reserves, unaware of the risk of fires around them. Increasingly, parks are closing on high fire-risk days, to reduce both the danger fires pose to people and the risk of people starting fires in the parks. Even so, large areas of reserves and parks remain accessible to the public.

Some of these visitors may be experienced outdoor hikers, very familiar with the pleasures and dangers of the Australian bush. Others may be city residents to whom bushfires are an occasional mention on the news. Many others may come from overseas where bushfires aren't an issue, and the risk does not even appear on their radar.

Last summer I went to stay in a bed-and-breakfast in the Otways. It was a beautiful timber house set on top of a ridge overlooking the national parks. It was also a stinking hot day, dry and windy, with temperatures tipping 40°C. It wasn't a day I would have chosen to be in the Otways or, indeed, away from home, but I had work to do and meetings scheduled, so I went anyway.

My hosts were kind and generous and suggested that, since the weather was so hot, it would be a good day to go into the forest and find a cool and shady spot. They made no mention of fire danger, no mention of the total fire ban. Among the reams of glossy tourism brochures and information sheets promoting the area, not one mentioned fire safety. I don't know what their personal bushfire plan was, or if they had one, but it obviously didn't include their guests.

It wasn't that my hosts didn't know about the fire risk. They were able to point out to me where the 1939 fires had come from, and the nearby road where four young children had been tragically killed as they fled up the road from their burning home. I guess talking about bushfires isn't all that great for business but when I looked through their guest book and saw how many of their visitors were from overseas—from places like Germany and Canada—I couldn't help wondering what duty of care we should have for visitors who stay in our homes and hotels, who visit our forests and beaches, who come to enjoy our relaxed and laid-back lifestyle.

Surveys of visitors along Victoria's Great Ocean Road suggest that, if there was a fire in the area, most people would get in their cars, either to drive away or, remarkably, to go and have a closer look. Only a small proportion would ask their hosts for advice and even fewer said they would seek advice from police or some other authority.

Fire safety is probably not what most visitors want to hear about when they are on holiday. We tend not to pay much attention to the safety messages on aircraft or to the evacuation maps on the back of hotel doors. That doesn't mean airlines and hotel operators shouldn't have disaster plans and communicate them to their customers. Only a few people have been killed by bushfires when travelling through an area, either for work or pleasure,[4] but the potential for worse fatalities is huge. Non-residents visiting bushfire areas are probably much less likely to be prepared. They are more likely to be unfamiliar with the area, unaware of safe or dangerous locations. They may be much less familiar with local fire history and risk. They are unlikely to be part of anyone's bushfire plan and are unlikely to have their own plan. More than 2 million people visit the parks along Victoria's Great Ocean Road every year. Even if those parks are closed, many visitors remain in the area, in the towns and on the roads, still very much at risk.

Lara 1969

As you drive through the wide, open paddocks on the multi-lane freeway from Melbourne to Geelong, you really wouldn't think you were at any risk from bushfire. Few trees break the view to the east or the west and only the conical shapes of long-dormant volcanoes break the flat line of the horizon as you speed around Port Phillip Bay.

Imagine then, what it must have been like for motorists on 8 January 1969, when a dense cloud of black smoke and embers descended on the roadway, cutting off all visibility and forcing a dozen or so cars to stop. Gale-force winds swept burning embers and debris over the cars. The fire had started the day before, when hot debris from a truck's exhaust started a grass fire on the edge of Bacchus Marsh Road near Anakie in western Victoria. The fire had been brought under control but reignited under the strong northerly winds, sweeping south towards the Princes Highway. Trapped in the unexpected dark and fiery heat, some people fled their cars in terror, perishing beside the road or in paddocks. Others died in their cars, with the doors open.

Some travellers survived, by huddling in their cars with their windows closed, and were able to drive off once the smoke lifted. While one car did catch fire, it still provided enough shelter for its occupants until the fire had passed, allowing them to escape safely. Seventeen people died in this unexpected tragedy on an open stretch of road with no obvious risk of fire.

The one place you don't want to be in a bushfire is out in the open. Probably the second-worst place to be is in a car. It is better than being out in the open, but only just.

Cars are not designed to withstand fire and, as car manufacturers have turned towards more flexible and lighter construction methods, cars have become less and less safe during fires. You might think that the most flammable part of a car is the petrol tank—which is true if the petrol spills out of the car in an accident, for example—but, sealed within an intact fuel tank, petrol is unlikely to explode without a spark to ignite it. Most tanks, including liquid petroleum gas tanks, will vent at high temperatures to release pressure and, while the ensuing flame-thrower reaction can look spectacular, it actually reduces the danger of the tank itself exploding. Petrol will spontaneously ignite if it gets hot enough, but only once temperatures exceed 200°C. At that temperature, the petrol tank exploding is the least of your concerns.

The things that do explode in car fires tend to be oil-filled cylinders, like the tail shaft and shock absorbers, but only once the car is well alight. For someone sheltering in a car during a fire, the greatest risk comes from plastic, glass and rubber. Starting from the ground up, we need to think about car tyres.

A neighbour of mine showed me a photo of a tractor burnt in the 2009 Strathewen fires. The front half of the old tractor was completely incinerated, while the back was perfectly intact and untouched. You wouldn't think there was much on an old tractor to burn, in the middle of an empty paddock, but the clue to its demise lies in its half-consumed state. My friend explained that the large back wheels of tractors are often half-filled with water, to give them added weight and stability. The water had prevented the back tyres from getting hot enough to ignite, just as you can boil water in a paper cup over a camp-fire without burning the cup. The tractor's air-filled front tyres had burst into flames, consuming its entire front half.

Car tyres are notoriously hard to extinguish. Once alight, tyre dumps can take days, weeks or even months to extinguish completely, exuding black clouds of toxic chemicals into the atmosphere. One dump fire in California took over two years to finally put out. Rubber may be slow to heat up, but it is also slow to cool down, so it frequently reignites, even when flames have been extinguished.

Tyres are not meant to be easy to burn. They require very

high temperatures before they will ignite, but their shape and location on a car probably makes them quite vulnerable to ember attack. There are lots of convenient nooks and crannies in and around the wheel arch and wheel itself for burning debris to lodge and gradually set fire to the wheels. Door seals are also common ignition points, for the same reasons.

All the plastic components of a modern car are flammable. Plastic burns really easily, even in the rain (which is useful to know if you ever need to light a signal fire from flotsam on a wet beach). The plastics used in cars tend to have lower ignition points than materials like rubber, so exterior plastic grilles and fittings catch fire relatively easily. As an added problem, the plastics inside the car may produce chemicals and fumes, making life inside even more difficult and hazardous.

Materials like aluminium, fibreglass and large expanses of glass may make for a racy lightweight car, but they do not provide the same fire protection as steel. At worst, glass can shatter and break in the heat; at best it offers very little in the way of thermal resistance, readily transferring any heat from the outside of the car to the inside. There is a good reason that the front seat of a car in the winter sun is a popular snoozing spot for grandfathers and cats the world over.

Once alight, many plastics and other synthetic materials are nigh-on impossible to put out with water alone. Firefighters put out lots of car fires—usually cars that have been stolen, dumped and deliberately lit—but they do it by smothering them with foam, not water. This raises the interesting dilemma of what happens during a bushfire when your car catches fire in the carport attached to your house. There were a number of cases in the recent Victorian bushfires where vehicles in carports were responsible for threatening or burning down houses. It is probably as well to regard cars in the same category as woodpiles and other flammable ignition points—either store them inside a sealed garage or well away from your house.

If you are caught in your car in a fire

The Bushfire Cooperative Research Centre guidelines for people in vehicles during bushfires include this safety advice:

- Park your car in the clearest possible area, away from any fuel (including grass and bush). If possible, park the car behind a cutting or outcrop that will also protect you from radiant heat.
- Close all the windows, doors and vents and turn the air conditioning off.
- Leave headlights and hazard lights on and engine running.
- Shelter below the window level under woollen blankets.
- Drink water to reduce dehydration.
- Stay in the car for as long as possible. If the heat feels worse than bad sunburn, it is too hot to go outside.
- If conditions in the car force you to leave (if it catches fire, if smoke or fumes become overpowering) try to find a patch of land that has already burnt and shelter there, with the woollen blanket.[5]

What if you are caught on the road in a car with no other option for shelter available? Notwithstanding all their inherent dangers, a car is still a safer place to be during a fire than outside in the heat. Obviously, if you know where the fire is and can turn around and drive away from it, that is the best thing to do. In reality, it is often difficult to see in smoky conditions, and spot fires and other fronts can appear from unexpected directions. Many people who die in cars during bushfires die from accidents when they run off the road or into other vehicles. Many people who think they are fleeing the flames end up driving right into them. If you do have to travel in fire-prone areas, keep your local radio on for any news and always carry a woollen blanket and water in your car.

It is possible that, if the people tragically killed on the Princes Highway all those years ago had stayed in their cars, they may well have survived. They had stopped in a relatively safe place—clear of vegetation under the cars and away from surrounding fuel. They only had to contend with the heat and embers borne on the wind and the cars may have been enough to protect them from those.

EVACUATION—A NOT-SO-SIMPLE SOLUTION

> **evacuate (verb) 1** *remove from a place of danger to a safer place.* **2** *leave (a dangerous place).*
>
> *Compact Oxford English Dictionary*, 2009

The difference between evacuation and leaving early is a subtle one. In effect, both involve the mass movement of people from a place of danger to a safer location. Evacuation carries with it a tone of authority, an overtone of enforcement. It is often something done by authorities to civilians. There is a sense of urgency, or requirement. The decision has been made by others and we are simply required to obey, for our own good. Leaving early, by contrast, seems a little wishy-washy. It leaves the decision firmly in the hands of those who may, or may not, be in danger. We must ourselves make the decisions; whether to go, when to go, where to go and what to take.

In fact, evacuation can be either mandatory or voluntary and even mandatory evacuation has a large element of choice. Few authorities have the resources or inclination to physically force unwilling residents to leave, even if they have the legal authority to do so. Evacuation is not so much a matter of forcing people to leave as clearly telling them when to leave.

In Australia, the national policy of leave early or prepare to stay and defend is a local modification of a concept more broadly recognised in international disaster responses as evacuation or shelter in place.[6] Which strategy is undertaken or promoted by authorities depends upon the nature of the threat and, importantly, the time frame in which it occurs.

Essentially, evacuation is only a safe option if the threat is recognised early enough to allow the safe and orderly movement of large numbers of people out of the area. Mass movement is hazardous at the best of times. Under conditions of stress and anxiety, it is even more dangerous. In a few cases, the risk attached to remaining in a particular place (say, a bomb threat or a structure fire) is greater than the risk of being caught in transit. In these cases, evacuation is always the best strategy.

In other cases, the risks of being caught in transit may be greater than the risks associated with remaining in place. For example, in certain industrial accidents, the safest option may be to evacuate, the next safest option is to remain inside, and the least safe option is to be caught in the open attempting to evacuate. In these cases evacuation only proceeds if there is enough time for people to completely leave the area before the threat actually materialises.

This is the scenario that applies to bushfires. In order to plan an effective evacuation strategy, we first need to consider how long it would take everyone in a given area to leave. The safe area needs to be well out of the way of any fire, so in most cases we would be planning for evacuations into urban areas.

Imagine the normal morning peak-hour rush. At these times our roads are probably operating at their maximum capacity. Roads are pretty much designed to meet this capacity—the smaller the load or population, the smaller the roads. There isn't much room for extra traffic without slowing down the flow considerably. In most Australian cities peak 'hour' is spread over two to three hours, to allow our current road network to carry everyone to work in the morning within a reasonable time. Now imagine doubling that load as all the non-workers, the people who work locally or in other areas, parents and children all add to the peak-hour traffic. Traffic would slow to a crawl. Even if numbers were managed to maintain flow (for example, by evacuating areas sequentially) it would still take hours to evacuate areas, even if everyone left in a prompt and orderly fashion.

And people do not leave in a prompt and orderly fashion. It takes time to evacuate. People are rarely prepared and, even when they are, they often underestimate the time it takes to catch the cat, dig out the photo albums, backup the computer or check on neighbours. It takes time for people to even decide that they are going to evacuate. Given an order to evacuate, very few people leap up and obediently leave the room immediately. They look around, see what everyone else is doing, consult, discuss and finally—eventually—grudgingly oblige, although only if the instruction continues to be given emphatically and repeatedly. This well-recognised element of human behaviour is rarely factored into evacuation models, even for buildings, yet it significantly reduces the effectiveness of evacuation.[7]

GRIDLOCK

A review of Australian road networks by the Transportation Research Board has already revealed that our roads are probably not capable of coping with mass evacuation and similar studies have been made in the USA.[8] If evacuation cannot be achieved in a timely manner, it is probably not the safest option. While evacuation might be a good option for those right on the urban fringe, for areas that have many roads and are in close proximity to safety, it might not be possible for those in more rural locations, with fewer and smaller roads and much further to travel, under potentially dangerous conditions, before they reach a safe location.

It is because of these time constraints, because of the dangers of being caught in transit and because of the unpredictability and speed of fires that most bushfire authorities only recommend evacuation (or leaving early) well before there is a fire in the area. It would be highly unsafe to evacuate once a fire had started.

Evacuation plans also assume that everyone faces the same risk as each other. A fit and healthy young farmer would be evacuated alongside an 80-year-old in a wheelchair. The resident of a brand-new brick home built and maintained to the highest fire safety standards would be evacuated alongside the resident of a rundown, overgrown, weatherboard cottage in the middle of the forest. The retired volunteer fire captain with 30 years of firefighting experience would be evacuated alongside the tree-changer suburbanite who had only just moved in last week. The risks fire poses to these people and their homes are not the same but they would all be evacuated.

Mandatory evacuation of everyone who lives in a bushfire risk area on every high fire-risk day would see the entire countryside and urban fringes vacated, possibly for weeks at a time. For many rural residents, this would simply not be feasible. Farms could not be run; businesses would go broke. This would place an unreasonable burden on many, both economic and emotional. American evacuations are not smooth military operations.[9] They cause great resentment, particularly when people are not able to return to their homes or when the threats turn out to be false alarms. Many people who have been evacuated in the past have said they will stay in the future.

If mass evacuation is not an option on all high fire-risk days, what about when fires threaten? This comes down to an issue of timing and accurate trigger points.[10] Given the speed at which fires travel and their unpredictability, it is unlikely that it would be possible to mass evacuate large numbers of people in advance of the fires. Our road networks are simply not big enough, and I suspect Australians generally are not compliant enough, for effective mass evacuation to work effectively and quickly enough.

Some American firefighters feel it is better to get everyone out of the fire zone, leaving firefighters free to concentrate on the fires rather than on people. But there is no doubt that having so many people on the roads also makes it difficult for emergency services to get to the fires. In addition, some argue that mandatory evacuation discourages people from making their homes fire safe, which makes it much harder and more dangerous for firefighters to save homes. As we know from fires in cities, one burning home will sequentially set fire to all the neighbouring homes, destroying whole towns and suburbs if they are not defended. Undefended houses are much more likely to catch fire and burn down than defended houses. This is precisely what happened in Canberra in 2003, where many of the houses were unattended, and in towns like Marysville, where the houses successively set fire to each other. Making sure your neighbours' house doesn't catch fire can be an important factor in protecting your own property. The greater the proportion of fire-safe houses, the lower the spread through built-up areas.

Evacuation places more people on the road, at the highest risk, and it increases property loss, by removing those who might be able to save their homes. In bushfires, homes can only be protected by people, whether these are volunteer firefighters or volunteer community members working to protect their own and their neighbours' properties. Taking everyone out of harm's way certainly keeps people safe but it comes at a price many are not prepared to pay. Whose responsibility is it to defend properties if the homeowners are not there to defend them? More volunteer firefighters? This seems a bit like robbing Peter to pay Paul.

RELEASE THE INNER FIREFIGHTER

With a bit of training and a bit of practice, plenty of advice and experience, any able-bodied person can become a volunteer firefighter. Wouldn't it be easier just to make sure we all had that training and support? Then we could at least each defend our own properties, leaving the volunteer firefighters to focus on strategic firefighting and defend those assets that have no-one else to defend them.

There are programs to provide this kind of training to members of the public. Some volunteer fire brigades allow anyone to complete their basic firefighter training on an 'obligation-free' basis. They argue that, after completing training, many people decide to continue volunteering, whether in a firefighting capacity or some other way. Even those who choose not to continue as firefighters return to their community better informed, better prepared and better able to promote fire preparedness to others in the community. It is a win-win situation for the brigades and the community.

With more support for training from central bureaucracies, the practice could be more widespread. It would cost money in the short term but it would save money, and lives, in the long term. It wouldn't suit everyone but it would certainly do no harm as one option among many. For such a strategy to be promoted, we would have to accept the notion that every able-bodied person has the potential to be a firefighter to some degree. This reality is already evidenced by the 200 000-odd volunteers who form the backbone of our bushfire fighting forces. These volunteers are drawn from all walks of life. They are ordinary men and women who, for a great variety of reasons, give up their time and effort to put out fires. For some reason, we still tend to think they are somehow different from the general community and there seems to be some kind of resistance to the idea that we could all do what they do (to a greater or lesser degree) if we just made the same commitment. Volunteer firefighters are members of the general community, not honorary career firefighters. They are just the same as you and me.

There are also many alternative systems to promote community awareness of bushfires. Community fire units in New South Wales provide limited training and equipment to community members to help them defend their own neighbourhoods, particularly on the urban fringe. In Victoria, community fireguard groups are established to facilitate fire preparation behaviour among neighbourhood groups, although without providing any specific hands-on training or expertise. These programs promote and support members of the public to become more engaged with fire protection, to learn and develop their fire planning and preparation, in concert with their neighbours. These are great programs that reach many people, but they are poorly funded, promoted and supported. For an annual cost of less than a single fire truck, they help many thousands of people to keep themselves and their houses safe in the event of a fire.

Some people haven't heard of these, and many others simply don't wish to. How to get more people interested and active in programs like these is one of the great challenges of bushfire safety across the country. It is not an insurmountable challenge. Twenty years ago, people who declined plastic shopping bags were regarded as slightly deranged, anyone who wore a fluorescent safety jacket was distinctly unmanly and the idea of wearing a helmet on a bike struck horror into any free-wheeling child's heart. Today, all these things are de rigueur—ordinary, everyday, unremarkable activities. Perhaps, one day, being prepared for bushfires will be normal too.

17 · SHELTER

Hide-and-seek is a universally favourite game among small children. Some hide over and over again in the same place, shrieking with excitement every time they are 'found'. Others master the art of hiding their eyes or head, neglecting to cover the most conspicuous parts of their body poking out from behind the couch. As they get older, children can get startlingly good at hide-and-seek, finding the most impossibly small places to tuck themselves away. Were it not for their inability to conceal their excitement with muffled giggles, they'd be nigh on impossible to find.

Yet, like most children's games, hide-and-seek conceals a far more sinister origin. Like all animals, we play in order to learn the skills we need to survive. Foals practise running and kicking, mountain goat kids try butting horns and climbing rocks, kittens play at pouncing and chasing. These young animals are learning how to be adults. Just as the early learning specialists tell us, we learn through play.

But what exactly are we learning? The foal is learning to outrun predators and to defend itself. The kid is learning how to establish dominance in a social hierarchy. The kitten is learning how to become a hunter.

Think again about those childhood games. When children play chasey in the school yard, they are learning to avoid predators. Dodging and weaving from the dreaded tag looks awfully like a classic cheetah-meets-gazelle scene from a wildlife documentary. And what about ball sports? Watch any young predator, particularly in a group, and you'll see this behaviour. It's the training ground of the hunter, tossing the long-dead remnants of a meal into the air and pretending it is still alive so that you can bring it to ground again.

The evolutionary history of humans is that we are both predators and prey and our past is written into our childhood games. We recapitulate our evolutionary development as we grow. Hide-and-seek is an ancient survival strategy—one in which humans are the prey. It is the hiding that is the fun bit (no-one likes being 'it'). When children hide, they are playing out some basic survival instinct, an ancestral memory as important for their survival as for any young gazelle on the savannah or deer in the forest.

Children instinctively respond to threats by hiding. If they are naughty, troubled or upset, they hide. If they are afraid, they hide. Children often continue hiding long after their mother's voice has gone from irritation to worry to near-hysterical distress. And children commonly hide when faced with a threat like fire—a potentially disastrous reaction.

Hiding in a cupboard, still and unmoving, might be the best option when faced with a fast-moving, movement-sensitive predator like a big cat. But survival strategies adapted for the savannah do not always stand us in good stead in the modern world. Children so often hide in house fires that firefighters are trained to search in cupboards and under beds if children are thought to be present. Adults, if they are awake, will run for the door. Children, alone and in the dark, will hide. Adult animals can run faster and further—running from a potential threat is a good idea. Most young animals can't outrun a predator, so they hide and hope to escape detection. Running away from a house fire is generally a good idea. Getting outside is the best and safest option. While adults tend to do this instinctively, children do not. They have to be taught to leave a building.

Running away from a bushfire is generally a bad idea (unless you do it well in advance). Getting inside as the fire front passes is the best and safest option. Yet adults do not seem to do this instinctively. They seek shelter as a last resort, in desperation, like a retreat to the childhood instinct to hide. While sheltering from a fire during its most intense period is essential, hiding is a different activity altogether. All too often, bushfires lead to house fires and, once a house catches fire, the hiding response is very dangerous indeed.

In the 1997 Dandenong Ranges fire, a young couple and their elderly neighbour sought shelter from a bushfire in the cellar of their house. As the house burnt around them, they were trapped beneath the burning building and tragically died in the small room they had imagined would be their refuge. In the Wangary fire in South Australia, one victim sheltered in her bathroom.[1] According to a firefighter who searched houses in Marysville after Black Saturday, around half of the fatalities were found sheltering in bathrooms.[2] The wettest room in the house seems to be a natural refuge for many people in the event of a fire, and yet it is not. Bathrooms, like internal cellars, rarely have an external exit. In the event of the house catching fire, people need to be able to get out and not be trapped inside with no escape.

House fires are massively hotter than most bushfires. Consider the density of fuel in a house, compared to the dispersed fuel in the same area of forest. Once alight, a house burns fiercely and intensely. People often comment on how hot bushfires are, citing the evidence of molten glass and metal. This particular heat does not come from the bush; it comes from the house itself. The bush may have set fire to the house but the house needs no further assistance to burn itself down.

Firefighters and researchers have a lot of experience

burning down houses. They do it for training and for research. While it is hard to measure what might happen in the intense heat of a bushfire, there is little evidence that houses suddenly explode into flames within seconds of a fire arriving.

The explosion seems to happen at that point in the combustion process where certain elements reach the right temperature to ignite themselves. Any bonfire, camp-fire or woodstove will demonstrate this process. Starting cold, the fire will smoke and smoulder, requiring much poking and prodding, stoking with kindling or other accelerants, and often a copious quantity of 'embers' in the form of matches. At some point, the fuel warms to a sufficient temperature for the electrons to jump about, releasing themselves into the atmosphere and combining with oxygen to create a sudden rush of fire.

The point at which this ignition is reached varies considerably for different fuels and different conditions but (for all except the most flammable fuels) the ignition point is reached at hundreds of degrees, temperatures that usually require sustained and direct flame contact rather than just ambient temperatures or radiant heat. Once the ignition point is reached, particularly in a vulnerable part of the house structure like a preheated roof space, the house suddenly appears to explode into flames. Firefighters often recognise these conditions—a telltale rippling in the air that signals a 'flash-over', as the gases almost reach ignition point.

In an extreme fire, it is difficult to know when that point is. In mild or moderate fire, the fire front typically passes by in five to ten minutes. You retreat inside when the heat is too much to take on exposed skin, and you go back outside to continue firefighting as soon as it is cool enough to tolerate. In an extreme fire, that danger period may be much longer—perhaps as long as an hour or more, depending upon the surrounding conditions. Among heavy fuel loads, or where wind prevents the heat from rising, the air itself may seem 'super heated' and too hot to breathe. Some locations may experience more than one front, coming from different directions as a fire burns around and up different sides of a hill, or changes direction with a cool change. Under these conditions, many houses need to be safer for longer if they are to protect people during the time when people are unable to protect their houses.

SAFE INSIDE

What if your house is not your refuge? What if the roof unexpectedly blows off in advance of the fire front, a window breaks or a tree falls on it, breaking the fragile protection the outer shell of the house provides against fire? What if you are in the middle of house renovations? Or if you slip on some water and sprain your ankle just as the fire front approaches? What if ... what if ...

By and large, houses protect most people from fires—but not always. Not all houses are defendable; not all houses are prepared. Sometimes fires are so fierce or the attacks so sustained that the fabric of the house is not able to withstand the heat, embers or even flames for long enough. Once a house is alight, there is only so long you can shelter in it before inside becomes more dangerous than outside. It is probably still worth staying inside for as long as possible.

Most houses catch fire well after the main fire front has passed. They don't usually catch fire from the front itself but from the embers that fall before and after the fire. As a consequence, by the time a house is fully alight and too dangerous to remain in, it may be safe to go outside and retreat to a clear or already burnt area. There are a number of provisos on this.

Firstly, it is not clear how long a house may remain safe. The increasing use of combustible, and toxic, substances like plastics may make smoke-filled houses uninhabitable much more quickly than researchers had previously thought. Secondly, the fire front may actually be several fronts, attacking a particular location at different times and from different directions. This is particularly a risk for hilltop properties, or during a wind change. Finally, how safe you are outside depends upon the surrounding fuel load. A grassy paddock will burn quickly and cool, making it a safe retreat very soon after the fire. A densely wooded forest will continue to burn, generating a lot of heat, even well after the fire has passed. Densely vegetated properties with no cleared areas may not have a safe outside refuge place in the event of the house burning down.

In all cases, additional protection is recommended. Woollen blankets have been a bushfire standard since the arrival of the first sheep. They remain one of the cheapest, most accessible and effective means of protection available to the average householder. Modern acrylic blankets may be super soft but wool can save your skin, quite literally.

Dams, riding arenas, paddocks, vegetable gardens and swimming pools have all been used as vital secondary refuges. They are not ideal primary refuges but they are important backups. Bear in mind that if you have to shelter in water, you still need to breathe. Should you find yourself in water while a fire front passes, make sure you can protect yourself from the super-heated air that will scorch airways and burn from the inside out. A wet woollen blanket is probably the best protection in a less-than-ideal scenario.

Water is not always the safest place to be. In the 1939 fires, some people sought shelter in metal rainwater tanks[3] but, surrounded by bushes and long grass or on top of wooden stands, there was little to distinguish a metal rainwater tank from a saucepan on a stove. Water transmits heat rapidly and deprives people of the ability to cool themselves down. Some died from the heat of the water and some drowned, too exhausted to climb back out again. Others were overcome by smoke.

Today, tanks are often made of plastic, as are above-ground swimming pools. While the water against the plastic will keep the temperature below melting or ignition point, any plastic above the waterline will melt in the heat or catch fire. Once this decomposition drops to the waterline, the water will start to escape, sometimes slowly, sometimes in a rush. At best, above-ground swimming pools provide a temporary refuge, provided there is no fuel close by. Dams and in-ground pools are a safer option, as a last resort but, again, this may depend on surrounding fuel and the level of protection from super-heated air.

DUGOUTS

Given that fire burns upwards, getting down low is always an advantage. Earth is a potent insulator against heat. Dugouts have long been a favoured refuge against bushfires and, despite their often primitive nature, they have proven their worth repeatedly. Ditches, drains, water culverts and even wombat holes have successfully sheltered people, provided they are large enough and deep enough and the entrance can be protected from burning fuels, smoke and embers.

The modern equivalent of the dugout is the bunker. This is not a cellar but a refuge located well away from the house. Typically bunkers are underground constructions but theoretically they could also be above ground, especially built to resist burning and heat. Bunkers are not designed to be the first point of refuge in a fire but rather the secondary refuge point if the house proves unsafe. They need to be far enough away from sources of fuel to be safe but close enough to the house to allow people to flee to them when required.

The ability of safely constructed dugouts, or bunkers, to save fives has been repeatedly demonstrated since at least 1926. The need for clearly mandated construction guidelines has been recognised since 1939.[4] And yet when fires struck Victoria in 2009 there were still no national standards for bunkers in Australia. In these fires, as in the past, homemade and makeshift bunkers saved many fives. A few did not. Homeowners have very little information on what bunkers should be like in order to be safe. We know that they need to be non-combustible, able to keep cool, exclude smoke and embers, and have entrances (preferably two) that cannot be blocked by falling trees or buildings. What we don't know is how different bunkers function in fires, how much air is required, what types of seals and materials work best, how far away from a house they need to be located.

Much of the official anxiety about bunkers stems from the horrific saturation bombing of civilians in Dresden by the Allied forces in World War II. People sheltered in their cellars as they had done during other bombing raids but these raids were of such intensity that they created a massive firestorm, incinerating and possibly suffocating as many as 40 000 people as they hid in their shelters. The fear is that an intense fire can suck the oxygen out of a shelter, leaving the occupants to suffocate.

We know so little about the conditions of truly severe fires that perhaps this is possible. But, in general, structure fires are hotter than bushfires and they last longer, so it seems unlikely that a bushfire would be as intense in one location as a city burning down. We need more research to understand if this is even likely to occur, or indeed physically possible. One interesting phenomenon that is observed in fires is the sucking of embers *into* a house. Often it seems that the embers are drawn into the house, perhaps due to a pressure differential. The movement of air could, at least, theoretically, occur in either direction.

The fact that 70 years after the Stretton Royal Commission highlighted the importance of bunkers, there were still no guidelines or recommendations, let alone standards for bunkers, is shameful. This failure simply reflects a lack of resources or effort on the part of successive federal governments and their respective agencies. We can send people safely to the moon, we can send them to the bottom of the ocean. We can even send men into burning buildings in special suits. There is nothing intrinsically difficult about testing bunkers and developing standards or guidelines for their construction that a bit of political will (and money) wouldn't fix. Our Commonwealth research agencies are equipped to do this work, just not funded to do it. In the scheme of things, the research itself is not that expensive. It just needs to be done.

On 24 September 2009, the Australian Building Codes Board announced that it would start work on a national standard for the design and construction of fire bunkers for personal use. The Board hopes to have the standards in place within six months.[5]

Some people argue that bunkers built to adequate safety standards will be far too expensive for most people to afford. Many people (including me) baulk at the cost of double glazing their windows in Australia, even though it offers substantial energy efficiencies, saves a lot of money in heating and cooling and offers substantial fire protection to one of the most vulnerable parts of a house. The argument that bunkers are too expensive is not one for fire agencies to decide. Expensive or not, householders deserve to have the information they need to make that decision for themselves. Otherwise some will spend the money anyway, on designs that may not offer the safety their owners hope for.

More important is the argument that bunkers may discourage people from defending their homes, from actively fighting fires, thus putting themselves at risk of a more severe fire than they need to. Or that people who should leave won't, because of the security a bunker provides. This is becoming a familiar argument: the risk that a backup option may discourage people from investing in the safest option. What we know from fires is that, for whatever reason or unforeseen circumstance, people aren't always able to follow the safest path. They don't prepare as much as they should, they don't leave when they should. Or they prepare very well but it isn't enough, or they intend to leave but are prevented from doing so. What we need to ensure is that the next-safest option is also available and that people aren't forced by lack of options to take the most risky options: fleeing in cars or, worse, on foot.

The advantage of bunkers is that they reduce the risk of people fleeing at the last minute, with their homes in flames. They give people a safe last resort. For those who choose to live in the highest risk areas, whatever the costs, that is surely worthwhile.

COMMUNITY REFUGES

Community refuges are another option when homes are not safe enough. After Ash Wednesday, there was a strong push for refuges, safe locations for people to flee to if their homes were threatened.

Community refuges are not the same as evacuation or emergency relief centres. Emergency relief centres are located outside a fire area. They are intended for people who have fled ahead of a disaster, or after a disaster. Most councils have plans in place to operate emergency evacuation centres, usually in local halls, basketball stadiums or civic centres. Community refuges are intended to operate during the fire and within the fire zone. They are like bunkers but able to hold more people. Both before and after Ash Wednesday, there was a move in Victoria and South Australia to identify safe community refuge areas, whether they were buildings, beaches or ovals. After Ash Wednesday, some councils and regional fire authorities identified and mapped the safest locations within their jurisdictions, essentially designating them as community refuges.

Some community refuges were better than others and there was little consistency applied to the process of designating them. The local oval or hall was identified in some areas. In other locations, facilities were specially constructed. Few refuges were even maintained with fire safety in mind, let alone developed, planned, built or clearly identified for this goal. Fears over liability, responsibility and cost paralysed local authorities. It seemed less risky to have no refuge than an unsafe one, and the concept of community refuges slowly and quietly slipped off the table.

Lack of funding and concerns over liability have seen most community refuges closed by local councils in Victoria. But just because there is no official refuge doesn't mean that people won't seek one in an emergency. Some residents will continue to believe that refuges still operate, despite advice to the contrary. In the event of a fire, they will turn up anyway. If the doors are locked, they will seek shelter somewhere else. The fire station at Kinglake West sheltered more than a hundred people on Black Saturday, simply because they had nowhere else to go.

The risk with designated community refuges is that people will use them as an evacuation point rather than a last resort. There is a great difference between designing a refuge intended for use by a hundred people and a point that might be used by an entire town. Perhaps the safer option is to ensure that there are multiple refuge points designated in a town. Bushfire researcher David Packham pointed out that there are already some buildings that are safer than others: often pubs and public toilets remain standing. Pubs probably have an advantage because a large number of residents congregate there and the more people present, the better the chances of defending the building against fire. Public toilets probably survive because of their proverbial 'brick dunny block' construction.[6] Packham's position echoes that of many bushfire researchers—we need buildings that double as shelters, not specialist shelters themselves.

In many communities, the buildings that house the most vulnerable people have been designated or utilised as refuges. In her seminal book on home bushfire safety, initially written in response to the Ash Wednesday fires, Joan Webster argues that many public buildings can and should be built to standards that allow them to operate as community refuges.[7] Any building in a risk area normally housing vulnerable members of the community needs to be capable of protecting them from bushfires: hospitals, aged-care facilities, schools, kindergartens. These buildings are often used by people who cannot be readily evacuated and they almost certainly will become centres to which people will flee or evacuate, whether they are designed for it or not. The Cockatoo kindergarten was a classic example in the Ash Wednesday fires; large numbers of people sheltered there when their own homes proved unsafe. The kindergarten's building design and environment kept it safe, with assistance from people wetting the building down and extinguishing embers.

The Victorian government policy of 'Neighbourhood Safer Places' looks set to resurrect the concept of community refuges under a new name but with all the same problems with implementation, funding, liability and practicality unresolved. In 25 years' time, will our neighbourhoods be any safer from bushfires? Or will we have, yet again, forgotten the lessons of our past?

I'm not convinced that we need 'neighbourhood safer places', although we do certainly need to make our neighbourhoods safer places to be during a fire. Imagine if, after Ash Wednesday, governments had made a requirement that all publicly funded buildings in high-risk bushfire areas be built and maintained to adequate bushfire standards. All kindergartens, schools, libraries, community centres, neighbourhood houses and sports clubs built over the last 25 years might then have been more likely to protect people from fires, instead of being, in many instances, potential death traps. Perhaps even more importantly, these building standards would have provided leadership in construction techniques, promoting and supporting the industry needed to provide the expertise and services to build to appropriate standards. Who knows, perhaps these raised standards might even have shifted the culture of bushfire preparation to a higher level, changing community expectations and flowing over into improvements in household building standards and maintenance. After all, if government departments will not spend the money to prepare their own buildings for bushfires, why would they expect private individuals to do so? Surely it is better to lead by example than to continue to spend money on messages people do not seem to want to hear.

Liability issues need to be resolved and funding needs to

be improved. There need to be many more buildings that are safer, not fewer. Perhaps funding could be available for renovations or innovations to improve fire safety for buildings like hotels or halls that might end up being used as fire refuges in the worst-case scenario. Perhaps buildings could be given fire-safety ratings, encouraging people to improve their own buildings and providing guidance as to which buildings are safest to shelter in. Many people in Kinglake survived because they knew which of their neighbours' houses was safest, because they knew where to retreat when their own house became unsafe. Many people in Marysville survived because they recognised the oval as a safe place. Many others did not know where to turn.

The issues involved in community refuges are complex but we have to base our actions not on what we would like people to do, but what they actually do. In a fire, many people leave late and seek shelter locally. We need to make sure they have somewhere to go.

SCHOOL REFUGES

Panton Hill was not burnt in the Ash Wednesday fires but lost five local men on their tanker. High community sensitivity to fire risk led to the building of a fire refuge at the school—but not without effort. It took 11 years of concerted lobbying to get funding from the Education Department for the building.

Panton Hill's fire refuge was finally constructed in 1993. The Education Department insisted that it must be square or rectangular (no returns), have a low, sloped, iron roof, be of brick construction, have metal shutters on all the windows and enclosed guttering. It has fire hoses and extinguishers. The architect managed to argue for a metal verandah, seats on a concrete base and small, double-glazed peephole windows to allow those inside to see what was happening outside. A proposed toilet was deemed an unnecessary expense—apparently 120 terrified primary school and kindergarten children could just hold on for the hour or two they needed to shelter inside. Or maybe use a bucket.

Victorian schools that have been identified as having low mains water pressure are required to install and maintain automatic water pumps and tanks that boost the local water supply in the event that the water pressure falls. These have to be funded out of the school's regular maintenance budget and are primarily intended as protection for building fires, rather than bushfires.

There doesn't seem to be any specific funding for schools in bushfire zones to build, improve or maintain fire-safety features. The costs of these initiatives (and they are considerable) must come out of the regular school budget or from the local community. Meanwhile an ever-increasing tangle of red tape makes it harder and harder for schools to keep their children, let alone their facilities, safe from bushfires. Health and safety regulations mean parents can no longer keep the gutters clean. Water restrictions have seen gardens dry to tinder boxes unless schools can afford expensive automatic watering systems.

Relatively few schools in bushfire areas have designated fire refuges. Their bushfire plans vary from having everyone clustering on the oval to sitting in the school library with blankets over their heads. Some schools have installed window shutters on old stone buildings but who is to say that the roof will stay on in a high wind, or that embers won't get under the eaves, or that the chimney is sealed? In Victoria at least, it is nearly impossible to get buildings inspected for fire safety, as there are no standards in place for fire refuges.

Few people seem willing to put their reputations on the line and declare that a particular building will keep a couple of hundred of our most precious and beloved bodies safe in the event of a disaster.

The cheapest option now seems likely to be the favoured one. Instead of making schools safer, it is easier to simply close them on high fire-risk days, sending the children back to homes that are likely to be more dangerous and less prepared. At least that way, any deaths are someone else's problem.

Schools are not the only locations where these concerns apply. There are many buildings that shelter the most vulnerable in our communities and many of these are funded, either in part or in total, from the public purse. Hospitals and medical centres tend to have good emergency plans in place for all sorts of disasters. But other facilities, particularly those that operate across private and public sectors, are less well prepared. Many aged-care facilities, child-care centres and pre-schools are in desperate need of better protection from bushfires. The very young and the very old are always at more risk from heat and stress. Evacuation, even when planned, is not always possible.

We have had this debate over bunkers and refuges so many times before. It was raised after 1939; it was raised after Ash Wednesday. After a brief flurry of activity and enthusiasm, well-intentioned recommendations gradually sink back into the distance, neglected and forgotten until the next disaster. As Jack McEwen said 70 years ago after the 1939 Canberra fires, if we do not learn from these events, we will just have to accept that 'nothing will ever be done'.[8]

18 · THE NEED TO KNOW

It is often said that we now live in an information age but humans are, and have probably always been, information junkies. We constantly seek, consume, utilise and discard information. Any way in which we obtain, manage and monitor this information is an inherently social activity.

How many decisions do you really make on your own? How often do you turn to your partner to decide what to have for tea? When you are asked a question you don't know the answer to, the first thing you do is turn to others for help. We are social decision-makers and at no time is this more obvious than in a disaster.

INFORMATION

There is often a shortage of information about where a fire exactly is, in local areas. Fires move rapidly; conditions change. Electricity may be cut; information from the fire ground needs to be circulated through central command posts before being publicly distributed, slowing down the information flow and reducing its level of local detail. Public broadcasters shoulder the greatest responsibility for accurate information, yet commercial broadcasters have the largest audience. Some fulfil their public responsibilities during the community's time of need. Others focus on their commercial imperatives.

Providing enough information, with local detail and when people need it, to help people make safe and informed decisions is one of the greatest challenges for fire authorities. We may not always seek out the information we need before a fire but it is the first thing we do when disaster strikes. Information is crucial for helping people to make the safest decisions they can, but delivering it during an emergency is fraught with difficulties. An informed community is a safer community.

The sea of peril

When I was a child, my parents decided to live on a boat. My grandparents were horrified. What a dangerous thing to do, taking a young family out on the high seas with storms and waves and the ever-present risk of drowning. At our second port of call, no more than a week after first setting sail, my mother was struck by a car in the street as she stood on the footpath outside a chemist shop. Her leg was broken and we spent the next year in one place, waiting for it to heal. My grandparents never mentioned the perils of sailing again.

Sailing certainly does carry risks—I recall drifting towards rocks, running aground, engine failures, minor collisions and major storms—but these risks can be managed and mitigated. Safety equipment, like life jackets, emergency beacons and flares, is a lifesaver when things go badly wrong but more important are the everyday safety precautions, like having a well-maintained boat and experienced crew. Even more important is safe behaviour. Watching the weather, both current and forecast, was ingrained in me at an early age. No-one was going to come by and tell us a south-westerly gale was headed our way and the anchorage we were in offered no protection. My family chose to live at sea—it was our responsibility to be aware of the risks and do our best to avoid them.

In 1998, the Sydney to Hobart yacht race ran into a storm. Sixty-six yachts retired from the race, 55 crew had to be rescued and six sailors died. The race was a disaster. Maybe the organisers should have cancelled or delayed the race. Maybe the participants should have withdrawn. Strict time schedules are often dangerous at sea. Keeping to a schedule doesn't allow you to leave port at the safest time, when the weather is favourable. On this occasion, the schedule required a fleet of flimsy, lightweight racing craft to set sail in the teeth of a gale.

Sudden, intense storms that develop from low-pressure systems in summer, like the one that hit the Sydney to Hobart field in 1998, are relatively rare. People knew the race was going to be a wild one but maybe no-one could have predicted just how wild it was going to be. But anyone could see the tightly packed isobars of the front heading across the Bight on the weather charts. A gale warning for the New South Wales south coast had been issued four hours before the race. It was upgraded to a storm warning just an hour after the race started—a full 20 hours before it hit the fleet.

Rather bizarrely, to my mind, much of the media attention on the investigation into these tragic deaths focused on the nature of the warnings given to the crews. Did they know they were sailing into a storm? More importantly, did they know what a 'storm' entailed? Much criticism was directed at the Bureau of Meteorology for failing to tell people how dangerous the storm might be and for not making more effort to contact the race organisers to make sure they knew how dangerous the storm was. The coroner concluded that, in addition to copious quantities of additional safety equipment, more information should be provided on maximum expected wave heights and wind speeds as well as averages. Only one recommendation out of 14 mentioned training of race participants.[1]

A gale is a gale, a warning is a warning and a storm is something you do your damnedest to avoid. If you don't know what they are, you shouldn't be sailing. We rightly place a great deal of emphasis on duty of care, on our responsibilities as a society, as organisations, as individuals, towards each other. Not many people talk about our responsibility for self-care, for looking after ourselves, for making safe decisions. We can't eliminate risk from our lives, even if we wanted to, but surely we need to take some personal responsibility for managing that risk, particularly when we have placed ourselves at risk by virtue of where we choose to live or what we choose to do.

WARNINGS

There seems little doubt that people want, and perhaps need, clearer instructions on when to leave and under what conditions to leave. Given the complexities of different levels of preparation and preparedness, combined with the vast differences in risk posed by different types of fires, this is not an easy demand to meet. It does not help that excessive bureaucratic fiddling with warning messages seems to promote vague ambiguity, instead of incisive clarity. Take, for example, Victorian authorities' advice in recent years to 'enact your bushfire plan' on a high fire-risk day.

This is actually an evacuation instruction but you could be forgiven for not recognising it. 'Enact your bushfire plan' is code for leave early or get ready to defend your home. Personally, I'm not that keen on opaque, waffly phrases, particularly in an emergency. They might satisfy all the different members of the committee debating the wording but they just don't work. If you want people to leave you need to say so. If you want people to get ready to stay and defend their homes, that is what you need to say. You need to say who the message is for, what the threat is and what people need to do. Warning messages need to be clear, specific, precise and comprehensive.[2] This hypothetical high fire-risk warning message provides an example of how they could be worded:

> Tomorrow is a day of total fire ban for the northern shire. People living in the high bushfire-risk areas of Johnsons Creek, Myrtlesea and Forestville should leave their homes now, unless they are fully prepared and equipped to fight bushfires, defend their homes and protect themselves from fire. If your home is not adequately prepared, or if you are not certain you are able to successfully protect yourself against fires, you should leave the area for a safe location away from high-risk areas until the area is declared safe to return.

Warning messages are not advertising jingles. They don't need to be short and snappy. American researchers at the Boulder Natural Hazards Centre have summarised a vast body of research on how warnings work and don't work.[3] This work, led by social sciences researcher Dennis Mileti, demonstrates how people respond to warnings and what type of warnings work best. But all too often emergency authorities get tangled up with marketing consultants who cater for a completely different audience in a very different state of mind. Authorities worry excessively about bad publicity. They worry that people might panic (even though there is no evidence that this happens). They worry that people will think they are the boy who cries wolf and stop listening to them.

These myths still dominate the minds of politicians and public servants, despite the fact that they have long been refuted. Panic is rarely a problem in an emergency—inaction is a far greater risk. False alarms might get a regular grumble and a whinge but, provided they are explained and are reasonable, they don't seem to lead to 'response fatigue'. People do ignore car alarms and faulty house fire alarms, but if you tell everyone the fire alarm has been fixed, and then it goes off, people will still file out of the building. Better still, get rid of the siren and have a real person make the announcement. Being open and honest, it seems, is a pretty good strategy.

Emergency services, as the vast majority of their staff know, are not about being popular. They are about being right. The evidence overwhelmingly shows that it is better to be safe than sorry.

WHO TO BELIEVE?

In the face of an impending threat, people seek out information. This doesn't mean they go to the library and do some intensive research or even that they will make a phone call to find out more. Information seeking does not have to be that active or deliberate. Rather, they will weigh up the information they have been given and assess it against a range of other sources of information. Does the threat conform to what we can see or hear? If you are told there is a damaging storm coming but outside it is a lovely, mild, sunny day, you are quite likely to discount this information because it doesn't conform to your own observations. If you are told that today's lovely, sunny weather will be pushed aside mid-afternoon by a violent storm, then your observations will confirm the information and you may be more likely to accept it as accurate. If you are also told about the low, dark line of clouds on the horizon to the south, you may just happen to look that way when you are out and the information will further be validated. Warnings need to be precise and accurate about what to expect.

The need for warnings to be precise and targeted was well illustrated with the fire that burnt Marysville in February 2009. A huge pillar of ominous, yellow-black smoke rose over the town well in advance of the fire but many people did not regard it as a threat. Even though general warnings were issued for Marysville, people did not put the warnings and the smoke together. The smoke plume was thought to come from the Kilmore fires or some other fire. Smoke very often does blow over towns from many, many kilometres away and represents no threat at all. People who live in forests are accustomed to the smoke of distant bushfires or the smoke of controlled burns. This smoke was different. Firefighters could see that it was different. The fire spotters could see that it was different. The residents of Marysville also needed to know that it was different. If they had seen that smoke cloud as a threat, it is possible that their response to the warnings would have been very different indeed. They might have left the area much earlier, or had a chance to find a safe place to shelter and defend when the flames arrived.

Authorities often assume that they are just that—*authorities*.

While it is true that information from official sources is important, people rarely just accept what any one authority says as gospel. Dennis Mileti's research suggests that warnings and information needs to come from a diversity of agencies and sources. There is no single, perfect authority to whom everyone will listen.

Having a senior commissioner deliver all the warnings may seem like a good idea but, in actual fact, people classify that as just one source of information (no matter how good it is) and automatically look for other sources for verification. This is why it is important to make sure that the different agencies and organisations all know what the key messages are and how to relay them. It is no good if the police tell people there is nothing to worry about, while the fire agencies are advising people to be on high alert. Consistent messages are important, although they can and should be delivered in different ways and forms.

Interestingly, Mileti's research suggests that if you must have one 'authority' figure, that figure should be a firefighter. Not the head of the firefighting agency in a crisply ironed uniform with epaulettes, speaking from a high-tech control centre, but a black-smudged, red-eyed firefighter holding a hose off the back of a truck. Firefighters rate highest of all as trustworthy sources of information.[4] People are probably more likely to listen to the advice of their local fire captain than they are someone from a central hierarchy. Local people often have standing within their community; they have local knowledge; they understand the conditions; and they know what they are talking about. These are the people we listen to, not someone far away in an air-conditioned office in the middle of the city. Accessing and utilising such local expertise is well known to be vital for fighting fires;[5] it is probably equally important in communicating with the public. Working out how to access that local expertise is a great challenge. The only thing we do know is that increasing centralisation of fire command structures is very unlikely to help.[6]

Informal sources are equally, if not more important in seeking information. The people whose opinions we really trust and rely upon are those of our friends and relatives. Even if we know, objectively, that other authorities may know better, we instinctively tend to follow what other people tell us to do.

Before, during and after a fire, like other disasters, we find out what friends and family know first, then we seek out information from authorities. We check websites, listen to the radio and turn on the television. When someone official makes an announcement we take note, and then go and check it. People rarely accept information at face value. They like to check a range of sources before making up their minds about its value. If different sources are providing different information, people may choose to heed one and ignore the other. Official information through the media is important but information given face to face is much more powerful. It may not matter how many pamphlets have been delivered, how much education is provided, how many carefully worded messages are articulated about not leaving too late in the event of a fire. When Fred puts his head over the back fence and shouts, 'The fire's coming, get in your car and get out of here!' a part of us wants to do as he says—even though we know it is a bad idea.

Warnings also need to be specific. It is all too easy for people to think that a warning does not apply to them and to ignore it. We are, after all, inherent optimists when it comes to personal safety. It may seem strange but, unless a warning is issued for a precise town, or suburb, people will believe that they are not at threat, even when warnings have been issued for all the surrounding districts. Perhaps just as important is to tell people when they are not at threat. Sometimes people in districts that are not at risk may feel threatened. Unnecessary evacuations or road traffic only adds to the chaos, so it important to tell these people what they can expect and what is expected of them. They need to be told that the smoke blowing over their suburb is not a threat, just as much as those who are at risk need to be told that the smoke foretells embers and flames.

The timing and nature of warnings is a highly contentious issue. In the days leading up to the Black Saturday fires, fire agencies and meteorologists provided graphic and grim warnings about the impending risks. These warnings were dutifully reiterated by politicians and journalists. The warnings could not have been worded more strongly or more explicitly. They weren't precise—they couldn't be, since no-one knew where a fire might strike—but they were clear and repeated across print, radio and screen. Surely no-one could say the community wasn't warned?

After Black Saturday, that is exactly what happened. Warnings, and a lack of precise warnings for particular communities and particular locations, dominated the press and the Royal Commission. Specific warnings, to particular townships, had not been issued or were not issued early enough. People had not been warned that they were under impending threat of the flames and their options for leaving or preparing were compromised. Many felt they lost their homes or lost loved ones because of this lack of warning. Communities felt let down.

Authorities felt aggrieved. They thought they had issued warnings. They issued them repeatedly in the days before the fire. They told people to 'enact their bushfire plan'—to leave early if they planned to leave or, if they were prepared, to get ready to defend their properties. The entire countryside should have been in a state of alert, with the vulnerable and at risk safely ensconced in the air-conditioned comfort of the cities and urban centres and the prepared actively patrolling their perfectly raked turf, hoses and pumps at the ready, waiting for the first sign of attack.

Sadly, they weren't. Not enough of them anyway. It was a classic mismatch between what the authorities thought they were saying and what the community heard. Few people heard the warnings issued as being specific to them. The warnings were too vague, not concrete enough for people to connect with. They didn't connect with many people's reality.

And when reality came knocking on the door, in the shape of a massive smoke plume, embers or even flames, the authorities were busy elsewhere, doing their best to fight the fires, keep their firefighters safe and protect whatever life and property they could. The trouble is that authorities can't warn people to leave once there is a fire in the area, because that is advising people to do something they know to be extremely dangerous. It might save their lives—but it might also send them to their deaths. That is an extraordinarily difficult call to make.

PREPARING THE UNPREPARED

If you are not ready and prepared to fight a fire by the time you can see embers, flame or smoke, you are not prepared at all and shouldn't be there. Fire authorities just don't know how to advise people who haven't heeded their messages, who aren't prepared or who are caught in a firestorm way beyond anyone's expectations or experiences. What can they say?

Community members are often warned of the need to have multiple contingency plans in place for fire. If plan A doesn't work, you need to consider plan B, or C or D or E. Perhaps the relevant authorities also need to have contingency plans for warning the community. It would be great if everyone went with plan A, but most don't, so we need to have plans B, C and D in place to fall back on. Where do people go when they are trapped and unprepared? What do we do with people who want to flee in the face of approaching flames? Where are the safest places in a community to shelter?

The trouble with backup plans is that you don't want people neglecting plan A, the safest and best option, because they think they can fall back on plan B. The worry is that having a plan B may weaken people's commitment to plan A. That is perhaps an irresolvable dilemma. But not having a plan B at all, particularly when we know that the majority of people don't follow plan A, just seems negligent. No wonder everyone feels let down.

V · THE FUTURE

19 · RECOVERY AND RESILIENCE

> *Look not too long in the face of the fire, Oh man. Never dream with thy hand on the helm! Turn not thy back to the compass; accept the first hint of the hitching tiller; believe not the artificial fire, when its redness makes all things look ghastly. To-morrow, in the natural sun, the skies will be bright; those who glared like devils in the forking fames, the morn will show in far other, at least gentler, relief; the glorious, golden, glad sun, the only true lamp—all others but liars!*
>
> Herman Melville, *Moby Dick*[1]

Driving along the road, you wouldn't even have known there'd been a fire. The messy tangle of trees and shrubs still crowded the narrow dirt road as it wound up towards the ridge. Cows and sheep grazed contentedly in the paddocks on either side, enjoying the first flush of green brought on by autumn rain. Vineyards tumbled down the hills in neat rows, laden with fruit ready for harvest. The landscape was a palette of gentle gum-tree greens and greys—soft, muted and reassuring.

Around a corner, over the brow of the hill, and we crossed a thin black line into devastation. As far as the eye could see, across the low hills and up onto the escarpment, the land was black and barren. Black fields stretched into endless forests of bare poles. There was not a blade of grass left on the ground, not a twig nor leaf on the trees. Here and there a rubble of pale bricks and twisted iron marked where a house once stood, where lives were once lived. Even the dams were black and still, coated in a thick murky layer of ash and cinders.

On the steeper, forested slopes, where the fire had raged the fiercest, the very soil itself had been burnt back to mineral earth. The thin, protective layer of leaf litter and mulch had disappeared and autumn rain now carved great eroded channels down the hills and into the gullies. The fire had raged from the floor of the forest all the way to the very tops of the trees, flames flinging themselves skyward above the canopy as it burnt. There was no canopy now. Rain fell between the bare poles to the earth below.

In some places, the fire had been less severe. The ground was burnt but the basal clumps of grass protruded with stubby fingers out of the ash. Some of the blady grass was even still green. The smooth-barked gums had been scorched around their bases, where piles of loose bark had been shed, but their trunks remained white and their leafy canopies turned orange and gold like the autumn leaves of deciduous trees. The thick, fibrous trunks of the stringybarks had candled, allowing the fires to smoulder higher into their canopies before eventually sizzling to a halt. But they too still had leaves—scorched, dry and dead leaves, but not burnt. Half-close your eyes and it looked like a northern hemisphere forest in autumn, ablaze with colour. But the smell of stale ash was overpowering, and the reds and golds like a pale echo of the fires that created them.

It was hard to believe that anything could recover after such an inferno. But the trees wasted no time mourning. The stringybarks were first, sending out pale pink coral sprays of new leaves from the buds beneath the bark on their trunks. Even trees on the most severely burnt slopes sent out new shoots, if not along their trunks, then from the underground lignotuber at their bases. The trees along the creek beds made rapid use of their access to water to regenerate and even the delicate, white-trunked gums blossomed with sprays of bright green growth.

The blackened stumps of tree ferns were soon topped with fresh fronds. Grass trees responded to the fire with indecent enthusiasm, sending forth the rampant, pale flower spikes beloved of bees and other pollinators. Bracken unfurled across the forest floor. Each shower of rain brought a carpet of regrowth to the fields and very soon they were no longer black but green.

Sometimes it is even hard to see where the fires have been.

THE LONG ROAD BACK

As the forest regenerates, survivors trickle back to their homes. Wallabies that once lived their secluded lives in the forests are now easy to spot as they emerge in search of food. Sharp-eyed birds scavenge and scratch through the cleared remains for an easy meal. People return to their homes to sift through the wreckage, shaking their heads over the melted remains of a car or the miraculous survival of a plastic swing set. Many blame the trees for the destruction but with each tree that reshoots, their spirits lift. Recovery seems a little more possible. As the forests recover, so too do the people who live there. They mourn their losses and grieve for those who have gone. They start to clean up, to think about rebuilding, to plan for the future.

For those who have lost friends and family to fires, recovery is a long and painful process.[2] Like a serious injury, the pain never entirely leaves and the scars remain visible. Life does not return to normal because normal is different now. Someone is missing from the fabric of our lives and nothing can mend that hole.

Trauma is not always associated with how much we have lost or what we experienced. Different experiences affect people differently. Some people have an incredible ability to bounce back from trauma, while for others a seemingly minor event can be the straw that breaks the camel's back. Just because someone else has suffered a greater loss does not mean that we do not grieve for losses that may seem trivial to others. There are no absolutes in grief, only relatives. The loss of a house maybe the loss of a lifelong dream; a dog may have been a cherished companion; a photo album may have been a last link to a life that has gone. People who weren't even in the fire zone can suffer from lasting trauma, depending on their personal circumstances.[3] Loss is not about numbers or quantities—no-one has a monopoly on grief—it is about how we feel, how we heal and how we recover.

REGAINING CONTROL

Fires, like many natural disasters, shake the foundations of our sense of control over our lives. Humans do not live with nature: we control it, or least try to. We build houses and office blocks to keep out the wind and rain. We heat them and cool them to our liking. We wrap ourselves in clothes. We seal the roads and footpaths to protect our feet from the elements beneath and erect shades to protect our heads from the elements above. We scurry from house to car to office and back again, only emerging into the sun when the weather is perfect. We live in a mollycoddled world of our own creation. Wind, rain and sun are reduced to mere inconveniences—slamming doors, flooding a porch or shrivelling the garden.

As much as we might be able to moderate the impacts of nature, we are only tinkering at the edges. We do not control nature at all and when nature, in any guise, exerts its superior force, we are left in shock and horror. Fires, tsunamis, cyclones, storms, earthquakes, landslides, floods and volcanoes rob us of our sense of control over our own lives, leaving us feeling powerless and afraid.

Control is central to recovery. The less prepared people are for fires, the greater their sense of lost control. The fire was unexpected; they didn't know what to do and they didn't know how to react. Their losses are attributed to an unstoppable catastrophe. Their survival is put down to a miracle.

People who are prepared tend to respond very differently to fires. For these people, fires are rarely completely unexpected. Fires are inevitable; it is just a question of when and where. Fires happen at certain times of the year, under certain conditions, during certain weather. They know which days are bad, which days are very bad and which days are downright horrors from hell. A fire can still take people by surprise, but they know what to do, they have a plan and, if plan A doesn't work, they move on to plan B, plan C, plan D, E or F. They don't have time to worry or freak out. They have a job to do and they do it. They survived because they were prepared and their losses are due to things they didn't plan for. Next time, they'll plan better.

Being prepared not only improves your chances of survival and of saving your home and property, it also seems to improve recovery. We may not be able to control fires, and even the best technology in the world is no match for a large fire. But being able to reduce the impact of a fire keeps us in control. We survived, not because of a whim of nature but because we were prepared. That sense of control is incredibly important.

Regaining control is a process that can occur in a variety of ways. Community involvement in rebuilding is often of great therapeutic value. In 2000 a prescribed burn in New Mexico swept out of the forest and across many Los Alamos communities. Despite the notion that the forest was responsible, in some sense, for the fires, locals organised volunteer groups to go into the forests to help clean up and replant. Rather than excluding people from the origin of the problem, they were encouraged to go in and see both the damage, and the regrowth, for themselves. Instead of being a distant ignition point, people could see what had happened, and also how it was possible to recover. One participant who lost her house explained, 'Seeing the forest recover helped me recover'.[4]

Regaining control is important enough for adults, but it is perhaps even more significant for children, whose sense of control over their own world is rather vulnerable at the best of times. In Los Alamos, schoolchildren were given the task of rebuilding an interpretative nature walk in the forest, helping them to understand what had happened and to make an active and valued contribution to recovery. They were able to share in both their own and their families' recovery, as well as the recovery of the environment.[5] After the Black Saturday fires, two Strathewen schoolchildren independently organised a photographic exhibition of bush regeneration, helping to raise funds, raise morale and, importantly, regain control over their community's future.

THE POWER OF NORMAL

After a disaster, we want to make things right, to put things back the way they were. It is certainly important to provide the support that affected communities need to rebuild. It is also important to ensure individuals and communities feel that they are the ones directing the rebuilding. The more opportunity they have for input into the process, the more effective their recovery is likely to be. We tend to see only the loss of homes and lives. We can't always replace them but we can help to restore people's sense of control. Often it is just the small everyday things that make the biggest difference to recovery. Getting back to a regular routine, eating home-cooked meals, catching up with family or going to the beach may be the best recovery activities around. Even just understanding that we have the capacity to heal ourselves can be incredibly empowering.

Rob Gordon has spent many years working on the front line of trauma psychology in Australia.[6] He argues that our social natures are crucial for understanding how we respond to and recover from disasters.[7] Humans are highly social animals. We define ourselves on the basis of our social networks. Under threat from fire, we rush to our neighbours to see what they are doing; we ring our families to check on them; we hurry to collect children or warn spouses to return from work. Our social networks are our defence against nature. They have protected humans for thousands of years in times of need and we instinctively strengthen them when we are under threat. In the era of modern communications, our networks buzz and crackle with renewed activity as the fire approaches.

But, when the fire front arrives, we are suddenly on our own. Phone and power supplies fail and we are too busy to use them anyway. In the dark and smoke, we can barely even see 5 metres away. It is nearly impossible even to check on others in the household, let alone know how the neighbours or friends are faring. We have only the knowledge of those connections to sustain us and the hope that they will still be there when the disaster is over.

Once the fire has passed, reconnecting these social links is our first priority. We check on friends and neighbours. We contact our families to reassure them that we are all right. Distant colleagues we'd all but forgotten about reappear to check on us. They are reassuring themselves but they are also reassuring us that they care, that we are still part of the fabric of society, that we have a place in a network of bonds and connections that are not broken, even when everything around us is destroyed.

When these bonds are not reconnected, when we remain isolated and dissociated from our friends and families, there is serious potential for long-term damage. Rebuilding and strengthening those social connections is incredibly important to recovery. Where circumstances keep people isolated after a disaster, and prevent them from reconnecting, the trauma can be profound. People feel abandoned, isolated and on their own and don't cope as well as those with a strong sense of community support. There is often not much anyone can do to help after a fire but sometimes the most important thing to do is nothing at all, just be there when you are needed.

Disaster psychologists have long recognised that communities affected by disaster need the opportunity to stay together.[8] After the devastation of Darwin by Cyclone Tracy in 1974, many residents were moved to other urban centres, like Brisbane. These people had little opportunity to share their experiences with others; their 'therapeutic community' was absent.[9] They had no-one to rebuild stories with and to help make sense of their experiences. Even though it is often difficult and raises complicated safety issues, keeping communities together and preferably close to their homes is now a priority in disaster recovery.

Talking is the primary means by which people make sense of what has happened to them. Fires are terrifyingly frightening experiences. They threaten our lives and we instinctively go into survival mode. Our bodies surge with adrenaline and we can be capable of extraordinary feats that we would normally find impossible. Our brains think differently too. Where we would normally weigh up the pros and cons of a situation before deciding on a course of action, in an emergency we bypass the normal executive control centres of the brain, allowing a more direct path between stimulus and response.

As a result of this survival-oriented brain activity, we don't lay down memories of events very effectively. Instead, what we remember tends to be sudden, vivid images. Other events and periods of time appear to be completely missing. We may recall a shed exploding into flames but have no recollection of how we got there. We may have a vivid memory of a person shouting from a car and be able to recall the features of their face in detail but not be able to remember where or when we saw them.

These images can be profoundly disturbing. They tend to recur in dreams or be prompted by unexpected events, like a loud noise or the smell of smoke. They often have a lot of emotion attached to them, emotion that also bubbles back to the surface with their reappearance.

Talking helps make sense of these isolated recollections and places them in context. By piecing together the events we can better understand what happened and how to deal with the emotions that are attached to those memories. The memory of the person in the car might be attached to fears about what happened to them. Finding out that they sheltered in a house down the street and are safe helps us to remove the fear from that memory. Rob Gordon believes that our anxieties are often driven by our fear of what might have happened, rather than what did happen. Focusing on the actual outcomes, rather than our fears of the worst outcomes, helps regain that sense of control that was damaged during the fire.

In some cases, of course, the worst did happen. Even in our deepest grief, we have a need to understand what happened, why and when. It may not heal the wound but it helps to staunch the bleeding, and gives us time to heal ourselves.

FRACTURES AND BONDS

Sometimes the bonds that reform after a fire are quite different from those that existed beforehand. People who barely knew their neighbours can be united by their common experiences. Priorities change. The demanding boss at work suddenly seems insignificant. People who had seemed busy and uninterested go to extraordinary lengths to help. Other friends find the tragedy all too difficult to deal with and move away.

As fire-affected communities rebuild, they reconstruct different social groupings.[10] Clusters of resentment or misunderstanding grow; a community united by grief and loss starts to fracture.[11] People who were prepared for the fire blame others who were not. Some blame the greenies, the council and the national parks for the trees. Others blame developers for building where they shouldn't, for building houses that weren't safe. They blame the authorities for not being able to stop the fire, for not being in the right place, for not giving them enough warning, for not keeping them safe. They blame other authorities for not burning off enough, for burning off too much and letting a fire get out of control. The insured resent the assistance given to the uninsured. Poorer neighbours feel neglected when richer neighbours can buy in the support they cannot access. Smaller towns feel larger towns get all the attention. Fracture lines form along pre-existing divisions in the community, creating even deeper divides that may take years to heal.[12] The Kilmore and Murrindindi fires fuelled a long-held community divide in surrounding areas, between development and conservation, leading to bitter feuds between those who would clear the surviving trees, and those who would save them.

Preventing these fractures can be difficult, particularly when there is so much rebuilding that needs to be done. It can be hard to attend to what seems like petty squabbles when deaths need to be investigated, water and power need to he restored, homes and schools rebuilt and whole townships restructured. Yet, like personal recovery, sometimes the solutions are relatively simple. Communication and information are always the key. People need to talk through their issues. This is how we work things through, how we make sense of what has happened.

While talking through the issues provides clarity and understanding, it can also reinforce misunderstandings and misinterpretations. Recognising where these misunderstandings are likely to occur and providing the information to deal with them can avoid some of these difficulties. The key to recovery lies in our own communities and allowing them to direct their own recovery, with support, understanding and assistance.

I heard someone ask, the other day, why, when we know that communities are so important in recovery, do we establish 'task-forces' and 'authorities' to oversee reconstruction?

It is a good question.

20 · WHAT NEXT?

Does our future lie in flames or is there the possibility of a future living with flames? The answer seems to lie along two different dimensions. The first is the nature of our changing climate: is our land likely to get hotter, drier and more fire prone in the future? The second dimension is a human one: are we capable of responding to the fire-prone nature of our landscape and finding safe ways of living with this inevitable risk?

CLIMATE CHANGE

The evidence for climate change on Earth is overwhelming, as incontrovertible as gravity, germ theory and genetic inheritance. The evidence is written in the very rocks beneath our feet. Ancient beaches in central Australia document the existence of a long-gone inland sea. Glaciers carved out the now-arid Flinders Ranges. The fossils of palms and cycads lie frozen in the rocks of Antarctica. The climate of the Earth has changed, is changing and will continue to change. End of story.

But the end of the story is where the questions begin. Climate change is not an exact science and the rocks themselves offer only an imprecise indication of time and temperature. A million years here or there means little in geological time. Biologically, a million years can encapsulate the evolution and extinction of whole lineages of species. Countless creatures must have appeared, proliferated and disappeared without leaving so much as a trace in the fossil record. Even the dramatic changes in climate in the past—planet-wide glaciation or 'snowball Earth'—can be difficult to trace, revealed only by the polished surfaces left by the ice, diamictites, tillites and dropstones left behind by glacial activity.[1]

How much the climate will change and how fast it will change are difficult questions to answer. More importantly, how much change can we, as a life form, as a species (much less as a culture or civilisation), survive? And how much have, or can, humans contribute to that change?

Since the evolution of modern animal life, the climate on Earth seems to have remained relatively stable, at least in comparison to earlier upheaval. The cycles of life, of oxygen and carbon, tend to stabilise the climate to conditions conducive for the continuation of life. But there is no guarantee that those conditions will remain conducive to our lives, to human life. In the course of the Earth's history, there have been five major mass extinctions, each of which wiped out up to 90 per cent of marine species (often the best-preserved ecosystems). Even relatively modest climate shifts can have profound ecological consequences for sensitive species.

We have already seen that fire plays a significant role in Australian ecosystems and has done for hundreds of thousands of years. We also know that the incidence of fire has increased dramatically in the last 40 000 years. It seems likely that this increase has been due to climate change, although it is possible that the dispersal of humans across the continent may have also played a role. Humans may also be responsible for the continuing high incidence of fire today, even though the climate should technically have returned to a wetter period. The long-term trend appears to be returning Australia to a condition similar to that in the last glacial age, when much of the land was desert.[2]

Cycles of life

For millennia the Earth warmed and cooled in cyclic fluctuations. Ice covered the Earth with frozen wastelands and giant glaciers. Snowball, or at least slushball, Earth, alternated with tropical poles. The growing proliferation of carbon-based life struggled against the fluctuating climate, clinging to the conditions that had allowed it to flourish. Salt-lovers retreated to shallow coastal bays and lakes as the seas sweetened. Heat-seeking bacteria clung to underwater volcanoes, revelling in their sulphurous plumes. Anaerobic bacteria burrowed deep underground or into the anoxic ocean depths, safe from the toxic oxygen that slowly built up across the planet.

Oxygen-loving creatures proliferated. As oxygen molecules surrounded the planet, ultraviolet light split them apart and recombined them into ozone, which diverted the most destructive components of the sun's radiation. The seas and lakes seethed with life, building great living reefs and underwater forests, filling the waters with swarms of creatures great and small. The lands were swathed with forests, luxuriant in the warm, moist atmosphere. Life created life.

Carbon cycled through earth and oceans, through plants and animals, through the atmosphere, and back to the earth again. Plants locked up the carbon. Death and fire released it. The conditions of life changed the climate and the climate changed the conditions for life in an eternal, interconnected cycle; constantly shifting, fluctuating and correcting, each in response to the other.

The risk of fire does not, of course, increase simply because it gets drier. There aren't many fires in the arid zones of Australia because there isn't much to burn. Deserts don't have a fire problem. What we do have to worry about is increasingly extreme cycles of weather. If wet becomes wetter and dry becomes drier, we have the potential both to increase our fuel loads and, simultaneously, the hot, dry conditions needed to burn them.

The Southern Oscillation that currently dominates Australia's weather, at least in the south-eastern part of the country, gives us just such a cycle of a few wet La Niña years followed by several dry El Niño years. This cycle creates the ideal circumstances for an increasingly fire-prone landscape. The wet years produce an abundance of vegetation, while the dry years provide the conditions to cure and burn it.

Predictive models of fire danger under different climate change scenarios by the CSIRO do suggest that the number of high fire-danger days will increase. Researchers modelled the effects of climate change under a variety of scenarios proposed by the Intergovernmental Panel on Climate Change. Given that these scenarios seem to be conservative, and that temperature changes appear to be happening more rapidly than might have been thought, it is probably best to use the worst-case scenario modelled.[3]

The outlook for 2050 differs for different climate zones. The south-eastern coastal cities and towns like Melbourne and Sale can expect their days of very high to extreme forest fire danger to increase from around 5 or 10 days per year to around 10 to 15. Coastal towns further north, like Sydney and Coffs Harbour, will increase their high fire-risk days to 12 to 17 days per year.

Inland, the fire risk increases dramatically. Towns like Canberra or Bendigo have around 7 to 20 high-risk days a year.

By 2050, they can expect around 20 to 35 days of very high risk. The increase is even more dramatic as the temperatures rise further inland. Wagga, Bourke and Mildura currently experience 50, 70 and 80 high-risk days per year. By 2050, these locations can expect two to three *months* of very high fire risk each year. Only in Tasmania is there no increase predicted, as it seems likely that increased humidity will offset increased temperatures.[4]

These models raise the potential of more fires, and more severe fires. But as we know, weather is only one factor entered into the fire equation. The other factor is fuel. Just as fire is self-sustaining, by creating the heat that sustains its own chemical reaction, fire is also self-limiting. The more often an area burns, the less it will burn, because there will be less fuel there to burn. These models do not take into account changes to the vegetation that might be induced either by climate change directly, or by reduced fuel loads. The weather in Mildura might make fires more risky, but the vegetation is unlikely to support them.

PREDICTING THE FUTURE FROM THE PAST

Attempting to understand patterns in Australia's fire risk that have occurred over hundreds of thousands of years, in order to see if things have changed in the last two centuries, is almost impossible. On the limited evidence we have available, it is not clear that the Black Saturday fires were any worse than those of 1851. The weather conditions were very similar; the fire descriptions are similar. The death toll was lower in 1851 but that could be explained entirely by population changes. Victoria's population has increased 45 times since 1851. Shortly before the gold rush, Victoria's entire population numbered just 77 000 people.[5] In 2009, nearly that many people lived just within the small area burnt by the February fires.

There is probably no way of knowing if fires are going to get worse in the near future. But just having fires as severe as they have been in the past is bad enough. Experience has shown that we are not equipped even for the conditions we currently know to occur, let alone any worse conditions. Why haven't we learnt from the fires in our past? We may not be able to change the weather—can we change our own behaviour to keep ourselves safer?

There have been a great many investigations, inquiries and studies into bushfires. In Victoria alone there have been two Royal Commissions, one Auditor-General's inquiry, a State Coroner's inquiry and five other major investigations into bushfires since 1939.[6] In addition, every incident is analysed, assessed and evaluated within each responsible agency for areas of improvement. Although every fire has its own unique circumstances and factors, the conclusions from each inquiry are remarkably similar. They all call for greater coordination and better organisation of firefighting efforts. They demand better fire prevention strategies and more fuel reduction burning. And they nearly always call for better community education and information.

Like any aspect of our fire history, there is an element of tedious repetition to these findings. Year after year, the same recommendations seem to be made, for better communication, better coordination, more local knowledge, more community education. So why, after so many inquiries and Royal Commissions, do we still have so many deaths from bushfires? Is there any reason to expect, given these past failures, that the Royal Commission into the Black Saturday fires will be likely to deliver any improvements on the past?

Having watched at least the initial stage of the Royal

Commission proceedings, I don't hold out much hope for any profound conclusions being reached. Lawyers, formidable though their skills are, are not specialist investigators, they are not specialist analysts and they are not specialist researchers—and they are very rarely bushfire experts. Put simplistically, our courts and legal system are intended to establish the guilt or innocence of a particular individual or party in relation to a particular crime. They do this within a framework of interrogation of witnesses. Identifying and rectifying systemic, and particularly social failures, is simply far too complex for this system. It is inevitable, despite the best intentions in the world, that if you ask a group of lawyers to investigate a problem, someone will end up being tried and found guilty. This may be psychologically satisfying for those who need someone to blame, but it does absolutely nothing to rectify the underlying problems that caused the events in the first place.

Despite the inherent limitations of various investigations, there have been many valuable and worthwhile recommendations made by past inquiries. Some of these have resulted in substantial improvements. There is no doubt that past inquiries have drastically improved the safety of firefighters, for example. Clearly problems have been identified and rectified, leading to a significant reduction in the number of firefighters being killed.

So why haven't the recommendations relating to the safety of the broader community succeeded in protecting them? Firefighter safety is typically the responsibility of one organisation. Identifying, implementing and monitoring improved practices, training and education within a single organisation is relatively easy and those changes are likely to be sustained by the institutional memory of the organisation. Community safety, on the other hand, is the responsibility of many.

First and foremost, community safety is the responsibility of individuals themselves. Yet getting the community as a whole, and those at highest risk in particular, to take responsibility for their own safety is notoriously difficult. We know perfectly well that poor diet, inadequate exercise and smoking are responsible for a large proportion of illnesses and deaths in our society. Yet how many of us eat as well (and as little) as we are supposed to? Or do enough exercise? How easy has it been to get people to give up (or not start) smoking?

It is easy to argue that if we are not prepared to take responsibility for our own well-being, then we must suffer the consequences. But, as social organisms, we share in each other's grief and hardship, at least to a certain extent. When someone is injured in a car accident, or a bushfire, it does not matter if their own actions contributed to that accident; we still take care of them. We don't deny people health services just because they have lived an unhealthy life.

Taking responsibility for individual safety is also a community responsibility, falling onto the shoulders of government and the organisations to which it delegates these responsibilities. The army keeps us safe from terrorists, invasions and other military risks. The police keep us safe from criminals. The doctors, nurses and hospitals take care of our health when we fall ill. A whole host of government agencies is responsible for keeping us safe from environmental pollution, dangerous goods, poor building design, shonky work practices and all sorts of other risks we face in everyday life. Such division of labour frees individuals within society to specialise in particular roles, without having to devote most of their time to basic survival tasks.

Where does community responsibility begin and individual responsibility end? In all these cases it is a question of degree. If a foreign army invades Australia, it is clearly up to the defence force to protect us. But if the war gets big enough, all able-bodied individuals will be called upon. Police protect us from crime but that doesn't mean we shouldn't lock car doors and keep an eye on our valuables. The difficulty with protecting the public from bushfires lies in the fact that it is a shared responsibility, between a relatively small proportion of the population who are at risk and a wide diversity of government agencies responsible for different elements of safety. While firefighting organisations in each state may seem the obvious responsible agencies, their ability to promote community safety may be limited by their focus on operational firefighting issues and the involvement of other agencies. Public land management departments may regulate fuel loads in some states, local councils may determine building and planning requirements in others, education departments dictate what happens in schools, recovery programs after fires are often coordinated by human services departments. And then there are competing interests that may work in a different direction from bushfire safety: conservation goals, industry groups, union interests, economic constraints and many others. In between all these different agencies and agendas there are a great many gaps into which bushfire safety recommendations can fall.

If we really want to reduce bushfire fatalities, we need a concerted and coordinated campaign across sectors to change people's behaviour. We need to tackle bushfire safety in the same way that we tackle the road toll, cigarette smoking, littering or sun safety. We need ongoing, constant, ever-changing but highly effective campaigns, designed to shift the perceptions of the community, to recognise that they are at risk and that they can do something to reduce their risk. We need to combine this with grassroots community education programs targeting those individuals at highest risk. We need to combine these campaigns with improved legislative structures for building and planning guidelines, to reduce the extreme risk cases, to prevent people from living in the highest risk areas with the least protection.

We need systemic change. We need processes that will continue to review and improve standards over the years in the absence of major fires. Many fire agencies already routinely investigate 'near misses' among their firefighters: accidents and incidents that had the potential to be fatal. By investigating these incidents, potential dangers and increasing trends can be identified and rectified before someone is killed. Hospitals increasingly operate similar systems, known as 'clinical governance', where certain indicators of quality are routinely investigated for systemic issues, allowing, for example, a sudden jump in infection rates through a changed cleaning regime to be corrected before it has major implications for patients. Perhaps we need a similar process for bushfire incidents in the community—a routine process for investigating house losses or property damage. Maybe this could even be a national system, allowing the best practices across states to be identified, then shared, promoted and adapted to the many different circumstances across the country. We don't need another round of recommendations that will be forgotten, underfunded or ignored within a few years or decades.

All these things cost money. There is no doubt that every life lost to a bushfire is a tragedy. So is every life lost to prostate or breast cancer—5000 to 6000 every year in Australia—and every life lost in traffic accidents, of which there are over 1000. With better screening, further research, more public awareness, better campaigns, many of these deaths are preventable.

Bushfire deaths are horrific, tragic and attract a great deal of public attention. But there are, on average, only five to ten deaths each year. And like traffic accidents, some of these deaths may not have been preventable no matter how much money was spent. A great many more people die in house fires each year than in bushfires. It costs money to save lives, so money tends to be spent where most lives can be saved.[7]

Bushfires do not threaten everyone. They threaten a small, well-defined group of people who live in high-risk areas. If you live within 500 metres of a substantial area of dry, native vegetation, then you belong to this group. I belong to this group. I choose to live in one of these areas. I choose to take that risk. I could choose not to—I could live somewhere much safer. The weight of responsibility for my choice rests largely with me.

The good news is that we probably do know what to do to keep ourselves safe. There were many houses, and shelters, that withstood extreme fire conditions on Black Saturday. Further research will undoubtedly identify and reconfirm the common features of the houses with the best survival prospects.[8] The newer the house and the more solidly built it is, the better it is likely to survive. Simple, well-sealed brick homes tend to survive better. Homes that are actively defended and well prepared are much more likely to survive than those that are not. Houses with well-positioned and protected pumps are more likely to survive.

We also know what puts us at risk. Homes close to vegetation and to continuous forests are at higher risk. The closer we wish to live to the bush, the safer our houses need to be. There is a clear linear relationship between proximity to vegetation and the security of our homes. Saving houses is not just about protecting property and reducing economic costs. Safe houses save lives. If our houses aren't safe then neither are our lives. We can keep ourselves safe either by distancing ourselves from the bush or by building safer homes. Either will do, but we have to do one or the other.

As a member of our society, I also think I should be entitled to a bit of support, even if I do make decisions that might not be the safest in the world. Just like motorbike riders or mountaineers or other mild risk-takers who are entitled to a bit of social support. We can't all play it completely safe all the time.

I'd like clearer guidelines about just how dangerous certain locations are. I'd like someone to assess my house for fire safety. I'd like readily accessible information on how different building materials cope in a fire. I'd like to know how to build a bunker. I'd like my children's school to have a regularly maintained fire-refuge. I'd like better building standards for fire zones. I'd like someone to tell my 80-year-old friends, living in an overgrown, weatherboard cottage in the forest, that they are highly likely to die if they stay during a fire. I'd like someone to know they are there and come to get them out.

But I can't have everything I might like. In the meantime, I'll just have to keep working on these things myself. It's the price I pay for living where I live. It's the Eden overhead. We live in a beautiful environment but it comes at a price. As Tim Flannery, a veteran of two bushfires (one of which claimed his house), once said, 'I'm beginning to think the gum trees may not love me as much as I love them'. But, like many Australians, he still lives in his home among the gum trees. Fire doesn't stop me from wanting to live where I live, but it is a risk I take very seriously.

There is only so much I can do to keep my family safe. It is my responsibility to do that, not someone else's, and I just have to hope that if disaster strikes, that I will have done enough.

ACKNOWLEDGEMENTS

I have always been fascinated by the role fire plays in the Australian bush and I have wanted to write a book on bushfires for many years. I am grateful to both Melbourne University Publishing and the Redmond Barry Fellowship (University of Melbourne and the State Library of Victoria) for providing me with the opportunity to realise this goal. The support of the staff and my fellow fellows at the State Library of Victoria has been much appreciated, and the library itself provided a much needed haven during the turmoil of the February fires, not to mention a safe refuge on high-risk days.

When I first began this project in July 2008, I had a very clear idea of what I wanted to write and how I wanted to write it. The fires of February 2009, which occurred so close to my own home, turned all those plans on their head and forced me to reconsider, not only this book, but also my own beliefs and preparations for bushfires. This book has ended up very different as a result.

Over the last year, I also worked as a researcher on psychological preparedness for bushfires for the Country Fire Authority. I would like to thank everyone I met during my time at the CFA for their friendliness and support, and for sharing their expertise and experiences so unstintingly. I couldn't list everyone individually but I do have to give special thanks to Alan Rhodes, Eli Niall, Wendi Fox, Ray Fritz, Angela Cook, Kim Stanley-Eyles and Helen Wositzky. While the research in this book is separate from my CFA work, my book could not help but be greatly enriched by the knowledge of my colleagues.

My time at the CFA also provided me with the opportunity to work with Rob Gordon, whose expertise and knowledge of recovery and trauma was hugely influential in shaping my understanding of individual and community reactions to disasters. I'm grateful for his generosity in sharing his experience with others.

Many people helped me with ideas, suggestions and resources for this book. I would like to thank Bill Wilson, retired Commissioner of Northern Territory Police and former volunteer firefighter, Julie Mitchell of Barcaldine Library and the Shannon family of Rodney Downs Station for their help with northern fires. Paul Smith, Paul Fowler, Lana Andrews, Zofie Lahodny-Gesco and Mark Gallagher from the Rural Fire Service of New South Wales provided valuable insights into New South Wales fires. Mike Ferdig and Cari Pukey from Orange County Fire Authority in California put our local experiences into a different perspective for me. Linden Gillbank, Cate Cousland and Siobhan Dee from the National Film and Sound Archive helped improve my historical accuracy.

I am very grateful to the team at Melbourne University Publishing for their encouragement of this project, particularly Tracy O'Shaughnessy and my various editors Cinzia Cavallaro, Eugenie Baulch and Susan Keogh, all of whom have been a delight to work with. I must also thank Alan Rhodes, Anne and John O'Brien, Mike Nicholls and Adrian Hyland for their patient proofreading and helpful comments.

This book is dedicated to all volunteer firefighters but particularly those at the Panton Hill Fire Brigade, with their steadfast refusal to take anything or anyone too seriously for too long—even their captain. It's lucky they have such an easygoing captain. Without his support and encouragement over the years, this book would never have been written. Thanks Mike, this one is for you.

NOTES

The Burning Bush

1 · Quoted in Hilverda S, *Hear the Siren: The First Fifty Years of the Panton Hill Volunteer Fire Brigade 1939–1990*, Panton Hill Volunteer Fire Brigade, Melbourne, 1990, p. 38.

2 · Luke RH, *Bush Fire Control in Australia*, Hodder & Stoughton, Melbourne, 1961.

3 · Emergency Management Australia, 'EMA Disasters Database', Attorney-General's Department, Canberra.

4 · Emergency Management Australia, 'EMA Disasters Database', Attorney-General's Department, Canberra.

5 · Stretton L, *Report of the Royal Commission to Inquire into the Causes of and Measures Taken to Prevent the Bush Fires of January, 1939, and to Protect Life and Property and the Measures to be Taken to Prevent Bush Fires in Victoria and to Protect Life and Property in the Event of Future Bush Fires*, State Government of Victoria, Melbourne, 1939, p. 4.

6 · Griffiths T, 'We Have Still Not Lived Long Enough', *Inside Story*, Australian Policy Online, 16 February 2009, viewed 10 May 2009, inside.org.au.

From Ashes to Ashes

1 · Singh G, Opdyke ND & Bowler JM, 'Late Cainozoic Stratigraphy, Paleomagnetic Chronology and Vegetational History from Lake George, NSW', *Journal of the Geological Society of Australia*, vol. 28, no. 3–4, 1981, pp. 435–52.

2 · Bowman DMJS, Tansley Review no. 101, 'The Impact of Aboriginal Burning on the Australian Biota', *New Phytologist*, vol. 140, 1998, pp. 385–410.

3 · Bowman DMJS, Tansley Review no. 101, 'The Impact of Aboriginal Burning on the Australian Biota', *New Phytologist*, vol. 140, 1998, pp. 385–410.

4 · Clode D, *Prehistoric Giants: The Megafauna of Australia*, Museum Victoria, Melbourne, 2009.

5 · Brain CK & Sillent A, 'Evidence from the Swartkrans Cave for the Earliest Use of Fire', *Nature*, vol. 336, no. 6198, 1988, pp. 464–6.

6 · Alperson-Afil N, 'Continual Fire-making by Hominins at Gesher Benot Ya'aqov,

Israel', *Quaternary Science Reviews*, vol. 27, no. 17–18, September 2008, pp. 1733–9.

7 · Bowler JM, Johnston H, Olley JM, Prescott JR, Roberts RG, Shawcross W & Spooner NA, 'New Ages for Human Occupation and Climatic Change at Lake Mungo, Australia', *Nature*, vol. 421, no. 6925, February 2003, pp. 837–40.

8 · White ME, *After the Greening*, Kangaroo Press, Sydney, 1994.

9 · Bowman, *New Phytologist*.

10 · Kershaw AP, 'Climatic Change and Aboriginal Burning in North-east Australia During the Last Two Glacial Interglacial Cycles', *Nature*, July 1986, vol. 322, no. 6074, pp. 47–9.

11 · Kershaw AP, Bretherton SC & van der Kaars S, 'A Complete Pollen Record of the Last 230 ka from Lynch's Crater, North-eastern Australia', *Palaeogeography Palaeoclimatology Palaeoecology*, vol. 251, no. 1, July 2007, pp. 23–45.

12 · Black MP, Mooney SD & Attenbrow V, 'Implications of a 14 200 year Contiguous Fire Record for Understanding Human-Climate Relationships at Goochs Swamp, New South Wales, Australia', *Holocene*, vol. 18, no. 3, May 2008, pp. 437–47.

13 · Black MP, Mooney SD & Attenbrow V, 'Implications of a 14 200 year Contiguous Fire Record for Understanding Human-Climate Relationships at Goochs Swamp, New South Wales, Australia', *Holocene*, vol. 18, no. 3, May 2008, pp. 437–47.

14 · Bowman, *New Phytologist*.

15 · Helgen KM, Wells RT, Kear BP, Gerdtz WR & Flannery TF, 'Ecological and Evolutionary Significance of Sizes of Giant Extinct Kangaroos', *Australian Journal of Zoology*, vol. 54, 2006, pp. 293–303.

16 · Miller G, Mangan J, Pollard D, Thompson S, Felzer B & Magee J, 'Sensitivity of the Australian Monsoon to Insolation and Vegetation: Implications for Human Impact on Continental Moisture Balance', *Geology*, vol. 33, no. 1, January 2005, pp. 65–8.

17 · Miller G, Mangan J, Pollard D, Thompson S, Felzer B & Magee J, 'Sensitivity of the Australian Monsoon to Insolation and Vegetation: Implications for Human Impact on Continental Moisture Balance', *Geology*, vol. 33, no. 1, January 2005, pp. 65–8.

18 · Latz P, *The Flaming Desert: Arid Australia—A Fire-shaped Landscape*, Peter Latz, Alice Springs, 2007.

19 · Clark RL, 'Pollen and Charcoal Evidence for the Effects of Aboriginal Burning on the Vegetation of Australia', *Archaeology in Oceania*, vol. 18,1983, pp. 32–7.

20 · Hughes PJ & Sullivan ME, 'Aboriginal Burning and Late Holocene Geomorphic Events in Eastern New South Wales', *Search*, vol. 12, no. 8, 1981, pp. 277–8.

21 · Schultz D (ed.), *Fire on the Savannas: Voices from the Landscape*, Tropical Savannas CRC, Darwin, 1988.

Fire-stick Farming

1 · Mitchell TL, *Journal of an Expedition into the Interior of Tropical Australia in Search of a Route from Sydney to the Gulf of Carpentaria*, Longman, Brown, Green, London, 1848, p. 249.

2 · Quoted in Flannery TF (ed.), *The Explorers*, Text Publishing, Melbourne, 1998, p. 30.

3 · Parkin R, *H.M. Bark* Endeavour: *Her Place in Australian History*, Miegunyah Press, Melbourne, 1997, pp. 216, 243.

4 · Mitchell, *Journal of an Expedition.*

5 · Blaxland G, *The Journal of Gregory Blaxland, 1813 Incorporating Journal of a Tour of Discovery across the Blue Mountains, New South Wales, in the Year 1813*, S. T. Leigh and Co., Sydney, 1813.

6 · Quoted in Benson JS & Redpath PA, 'The Nature of Pre-European Native Vegetation in South-eastern Australia; A Critique of Ryan, D. G., Ryan, J. R. and Starr, B. J. (1995) *The Australian Landscape—Observations of Explorers and Early Settlers*', *Cunninghamia*, vol. 5, no. 2, 1997, p. 310.

7 · Quoted in Flannery (ed.), *The Explorers*, pp. 25–6.

8 · White J, *Journal of a Voyage to New South Wales*, J. Debrett, London, 1790, pp. 150–1.

9 · Vancouver G, *A Voyage of Discovery to the North Pacific and Round the World Performed in the Years 1790, 1791, 1792, 1793, 1794 and 1795 in the* Discovery, 2nd edn, John Stockdale, London, 1801.

10 · Hunter J, to the Duke of Portland, 10 June 1797, in *Historical Records of New South Wales*, 1797, vol. 3, p. 220.

11 · Parkin, *H.M. Bark* Endeavour, pp. 367–70.

12 · Peron F & de Freycinet L, *Voyage of Discovery to the Southern Lands*, Friends of the State Library of South Australia, Adelaide, 2006, p. 190.

13 · Peron F & de Freycinet L, *Voyage of Discovery to the Southern Lands*, Friends of the State Library of South Australia, Adelaide, 2006, p. 194.

14 · Mitchell, *Journal of an Expedition*, p. 297.

15 · Macgillivray J, *Narrative of the Voyage of HMS* Rattlesnake, *Commanded by the Late Captain Owen Stanley RNRFS &c During the Years 1846–1850 Including Discoveries and Surveys in New Guinea, The Louisiade Archipelago, etc. to Which Were Added the Account of Mr. E. B. Kennedy's Expedition for the Exploration of the Cape York Peninsula*, T. & W. Boone, London, 1852.

16 · Quoted in Flannery (ed.). *The Explorers*, pp. 55–6.

17 · Giles E, *Australia Twice Traversed: The Romance of Exploration, Being a Narrative Compiled from the Journals of Five Exploring Expeditions into*

and through Central Australia and Western Australia from 1872 to 1876, Sampson Low, Marston, Searle & Rivington, London, 1889, p. 111.

18 · Carnegie D, *Spinifex and Sand: A Narrative of Five Years' Pioneering and Exploration in Western Australia,* C. Arthur Pearson, London, 1898.

19 · Plomely NJB, *Friendly Mission: The Tasmanian Journals and Papers of George Augustus Robinson 1829–1834,* Tasmanian Historical Research Association, Hobart, 1966, p. 840.

20 · Stokes JL, *Discoveries in Australia with an Account of the Coasts and Rivers Explored and Surveyed During the Voyage of HMS* Beagle *in the Years 1837–38–39–40–41–42–43,* T. & W. Boone, London, 1846, p. 228.

21 · Plomely, *Friendly Mission.*

22 · Jones R, 'Fire-stick Farming', *Australian Natural History,* September 1969, pp. 224–8.

23 · Flinders M, *Narrative of the Expedition of the Colonial Sloop* Norfold, *7 October 1798–12 January 1799,* Matthew Flinders Archive, 1798–99, Mitchell Library, Sydney.

24 · Antoine Bruni D'Entrecasteaux in Duyker E & Duyker M, *Bruny d'Entrecasteaux: Voyage to Australia and the Pacific 1791–1793,* Melbourne University Press, Melbourne, 2001.

25 · Balmer J, Botanist, Tasmanian Parks and Wildlife Service, Department of Primary Industry, Water and Environment, personal communication, 1998.

26 · National Parks and Wildlife Service, *Wollemi Pine Recovery Plan,* New South Wales National Parks and Wildlife Service, Sydney, 1998.

27 · Leichhardt L & von Neumayer G, *Dr. Ludwig Leichhardt's Letters, from Australia, During the Years March 23, 1842, to April 3, 1848: with an Appendix: Dr. Ludwig Leichhardt as Naturalist and Explorer,* Pan Publishers, Melbourne, 1945, p. 36.

28 · Mitchell, *Journal of an Expedition,* p. 170.

29 · Eyre EJ, *Journals of Expeditions of Discovery into Central Australia and Overland from Adelaide to King George's Sound in the Years 1840–1: Sent by the Colonists of South Australia, with the Sanction and Support of the Government: Including an Account of the Manners and Customs of the Aborigines and the State of Their Relations with Europeans,* T. & W. Boone, London, 1845, p. 140.

30 · Abbot I, 'Aboriginal Fire Regimes in South Western Australia: Evidence from Historical Document', in Abbott I & Burrows N (eds), *Fire in Ecosystems of South West Western Australia: Impacts and Management,* Backhuys, Leiden, 2003, pp. 119–46.

31 · Keppel H, *A Visit to the Indian Archipelago in HM Ship* Meander, Richard Bentley, London, 1853, p. 154.

32 · Curr E, *Recollections of Squatting in Victoria 1841–1851*, Robertson, Melbourne, 1883, p. 87.

33 · Mitchell, *Journal of an Expedition.*

A Walk in the Park

1 · Meredith L, *My Home in Tasmania During a Residence of Nine Years*, John Murray, London, 1852, pp. 107–9.

2 · Diamond J, *Guns, Germs and Steel: A Short History of Everybody for the Last 13,000 Years*, Vintage, London, 1997, p. 125.

3 · Parkin R, *H.M. Bark* Endeavour: *Her Place in Australian History*, Miegunyah Press, Melbourne, 1997, p. 189.

4 · Quoted in Flannery TF (ed.), *The Explorers*, Text Publishing, Melbourne, 1998, pp. 164–5.

5 · White J, *Journal of a Voyage to New South Wales*, J. Debrett, London, 1790, p. 116.

6 · Dumont d'Urville JSC, *An Account in Two Volumes of Two Voyages to the South Seas*, Melbourne University Press, Melbourne, 1987, p. 56.

7 · Mitchell TL, *Journal of an Expedition into the Interior of Tropical Australia in Search of a Route from Sydney to the Gulf of Carpentaria*, Longman, Brown, Green, London, 1848.

8 · Mitchell TL, *Journal of an Expedition into the Interior of Tropical Australia in Search of a Route from Sydney to the Gulf of Carpentaria*, Longman, Brown, Green, London, 1848.

9 · Benson JS & Redpath PA, 'The Nature of Pre-European Native Vegetation in South-eastern Australia: A Critique of Ryan, D. G., Ryan, J. R. and Starr, B. J. (1995) *The Australian Landscape—Observations of Explorers and Early Settlers*', *Cunninghamia*, vol. 5, no. 2, 1997, pp. 285–328.

10 · Howitt AW, 'The Eucalypts of Gippsland', *Transactions of the Royal Society of Victoria*, vol. 2, no. 1, 1890, pp. 81–120.

11 · Howitt, *Transactions of the Royal Society*, p. 111.

12 · Griffiths T, *Forests of Ash: An Environmental History*, Cambridge University Press, Melbourne, 2001, p. 27.

13 · Walker JB, 'Some Notes of the Tribal Divisions of the Aborigines of Tasmania', *Early Tasmania: Papers Read before the Royal Society of Tasmania During the Years 1888 to 1899*, M. C. Reed, Hobart, 1989, p. 268.

14 · King PP, *Narrative of a Survey of the Intertropical and Western Coasts of Australia*, John Murray, London, 1827, p. 17.

15 · Webster J, *The Complete Australian Bushfire Book*, Penguin Books, Melbourne, 1989.

16 · Ward DJ, Lamont BB & Burrows CL, 'Grasstrees Reveal Contrasting

Fire Regimes in Eucalypt Forest before and after European Settlement of Southwestern Australia', *Forest Ecology and Management*, vol. 150, 2001, pp. 323–9.

17 · Zylstra P, *Fire History of the Australian Alps: Prehistory to 2003*, Australian Alps Liaison Committee, Sydney, 2006.

18 · Zylstra P, *Fire History of the Australian Alps: Prehistory to 2003*, Australian Alps Liaison Committee, Sydney, 2006.

19 · Lunt ID, 'Two Hundred Years of Land Use and Vegetation Change in a Remnant Coastal Woodland in Southern Australia', *Australian Journal of Botany*, vol. 46, no. 5–6, 1998, pp. 629–47.

A Growing Risk

1 · Sentinel home page, Geoscience Australia, Canberra, viewed 7 July 2009, http://sentinel.ga.gov.au/.

2 · Loane IT & Gould JS, *Aerial Suppression of Bushfires: Cost–Benefit Study for Victoria*, National Bushfire Research Unit, CSIRO, Canberra, 1986.

3 · Orr E, *My First Visit to the Bush in Victoria Australia and Recollections of Black Thursday, the Great Bushfire of 1851*, State Library of Victoria, Melbourne, 1890, p. 5.

4 · Orr E, *My First Visit to the Bush in Victoria Australia and Recollections of Black Thursday, the Great Bushfire of 1851*, State Library of Victoria, Melbourne, 1890, p. 5.

5 · Orr E, *My First Visit to the Bush in Victoria Australia and Recollections of Black Thursday, the Great Bushfire of 1851*, State Library of Victoria, Melbourne, 1890, p. 5.

6 · Howitt W, 'Black Thursday—The Great Bush Fire of Victoria', *Cassell's Illustrated Family Paper*, vol. 1, no. 6, 1854.

7 · Emergency Management Australia, 'EMA Disasters Database', Attorney-General's Department, Canberra.

8 · Anon., 'Bushfires of 1851, Black Thursday, Disaster Recalled', *Argus Newspaper*, 20 February 1926.

9 · Edwards DH, *Diamond Valley Story*, Shire of Diamond Valley, Melbourne, 1979, p. 97.

10 · Finn E, *Chronicles of Early Melbourne, 1835 to 1852: Historical, Anecdotal and Personal*, Fergusson & Mitchell, Melbourne, 1888.

11 · Howard ECO, 'Black Thursday', *Argus*, 28 June 1924, p. 6.

12 · *Launceston Examiner*, 8 February 1851.

13 · Anon., 'The Black Forest of Victoria', *Australasian*, vol. 3, no. 1, April 1851, pp. 456–72.

14 · Westgarth W, *Personal Recollections of Early Melbourne and Victoria*, Robertson, Melbourne, 1888.

Fire in the Outback

1 · Hoch I, *Barcaldine 1846–1986*, Barcaldine Shire Council, Barcaldine, 1986.

2 · Quoted in Hoch I, *Barcaldine 1846–1986*, Barcaldine Shire Council, Barcaldine, 1986, p. 57.

3 · Hoch I, *Barcaldine 1846–1986*, Barcaldine Shire Council, Barcaldine, 1986, p. 65.

4 · *Argus*, 3 August 1914.

5 · Hoch, *Barcaldine*, p. 67.

6 · Barcaldine Historical Society, *Barcaldine Historical Society Inc. Commemorates the Disastrous Bush Fires of Saltern Creek, Willoughby and Rodney Downs*, Barcaldine Historical Society, Barcaldine, c. 1980.

7 · Barcaldine Historical Society, *Barcaldine Historical Society Inc. Commemorates the Disastrous Bush Fires of Saltern Creek, Willoughby and Rodney Downs*, Barcaldine Historical Society, Barcaldine, c. 1980, p. 8.

8 · Barcaldine Historical Society, *Barcaldine Historical Society Inc. Commemorates the Disastrous Bush Fires of Saltern Creek, Willoughby and Rodney Downs*, Barcaldine Historical Society, Barcaldine, c. 1980.

9 · Mannix I, *Great Australian Bushfire Stories*, ABC Books, Sydney, 2008.

10 · Hoch, *Barcaldine*, p. 68.

11 · Quoted in Fahy M, *The Day the Flames Came: Dwellingup 1961*, DVD, 30 minutes, CALM, FESA and the Bushfire CRC, Australia, 2006.

12 · Schapel AE, *Inquest into the Deaths of Star Ellen Borlase, Jack Morley Borlase, Helen Kald Castle, Judith Maud Griffith, Jody Maria Kay, Graham Joseph Russell, Zoe Russell-Kay, Trent Alan Murnane and Neil George Richardson*, Coroner's Court of South Australia, Adelaide, 2007.

13 · Latz P, *The Flaming Desert: Arid Australia—A Fire-shaped Landscape*, Peter Latz, Alice Springs, 2007.

14 · Morehouse BJ & Orr BJ, 'Integrating Science and Community into Decision Support', *Wildfire Risk: Human Perceptions and Management Implications*, Resources for the Future, Washington, 2008, p. 196.

15 · Mitchell JK, 'Natural Disasters in the Context of Mega-cities', *Crucibles of Hazard: Mega-cities and Disasters in Transition*, United Nations University Press, Tokyo, 1999, p. 18.

Living in the Forest

1 · 'Victoria Scorched by Worst Bushfires in State's History', Movietone News, vol. 10, no. 5, 21 January 1939, Thought Equity Motion, A0235, National Film and Sound Archive.

2 · Hawkless-Consulting, *Regulatory Impact Statement: Rural Fires Regulation 2008*, Rural Fire Service, Sydney, 2008, p. 12.

3 · Noble WS, *Ordeal by Fire: The Week a State Burned Up*, Melbourne, Jenkin Buxton Printers, 1977.

4 · Griffiths T, *Forests of Ash: An Environmental History*, Cambridge University Press, Melbourne, 2001.

5 · Stretton L, *Report of the Royal Commission to Inquire into the Causes of and Measures Taken to Prevent the Bush Fires of January, 1939, and to Protect Life and Property and the Measures to be Taken to Prevent Bush Fires in Victoria and to Protect Life and Property in the Event of Future Bush Fires*, State Government of Victoria, Melbourne, 1939.

6 · McEwen J, *Canberra Times*, 23 January 1939, p. 2.

7 · Lamont BB, Ralph CS & Christensen PES, 'Mycophagous Marsupials as Dispersal Agents for Ectomycorrhizal Fungi on *Eucalyptus calophylla* and *Gastrolobium bilobum*', *New Phytologist*, vol. 101, no. 4, 1985, pp. 651–6.

8 · Fahy M, *The Day the Flames Came: Dwellingup 1961*, DVD, 30 minutes, CALM, FESA and the Bushfire CRC, Australia, 2006.

9 · Fahy M, *The Day the Flames Came: Dwellingup 1961*, DVD, 30 minutes, CALM, FESA and the Bushfire CRC, Australia, 2006.

Urban Impact

1 · Kumagai Y & Nojima Y, 'Urbanization and Disaster Mitigation in Tokyo', *Crucibles of Hazard: Mega-cities and Disasters in Transition*, United Nations University Press, Tokyo, 1999, pp. 56–91.

2 · Fahy M, *Black Tuesday: 1967 Tasmania Bushfires*, DVD, 29 minutes, Bushfire CRC, Australia, 2005.

3 · Fahy M, *Black Tuesday: 1967 Tasmania Bushfires*, DVD, 29 minutes, Bushfire CRC, Australia, 2005.

4 · Franklin E, 'Elizabeth Franklin Remembers the 1994 Bush Fires', *Local Studies—Western Foreshores*, Pittwater Library Services, Sydney, 1994.

5 · *Canberra Times*, 16 January 1939, p. 1.

6 · Mannix I, *Great Australian Bushfire Stories, ABC Books, Sydney*, p. 10.

7 · Attewell F, 'California Police Shoot Dead Suspected Arsonist', *Guardian*, 25 October 2007.

The Big Ones

1 · Emergency Management Australia, 'EMA Disasters Database', Attorney-General's Department, Canberra, 2009.

2 · Country Fire Authority, *The Major Fires Originating 16th February, 1983*, Country Fire Authority, Melbourne, 1983.

3 · Nicol M, 5DN live radio broadcast, Adelaide, 16 February 1983.

4 · Milligan J, *Ash Wednesday in Upper Beaconsfield*, Pakenham Gazette, Pakenham, 1992, p. 18.

5 · Country Fire Authority, *The Major Fires.*

6 · Milligan, *Ash Wednesday*, p. 35.

7 · Hilverda S, *Hear the Siren: The First Fifty Years of the Panton Hill Volunteer Fire Brigade 1939–1990*, Panton Hill Volunteer Fire Brigade, Melbourne, 1990, p. 66.

8 · Krusel N & Petris SN, *A Study of Civilian Deaths in the 1983 Ash Wednesday Bushfires Victoria, Australia*, Country Fire Authority, Melbourne, 1992.

9 · Willans A, 15 June 2009, in Teague B, McLeod R & Pascoe S, *Transcript of Proceedings, 2009 Victorian Bushfires Royal Commission*, CRS Wordwave Pty Ltd, Melbourne, 2009, p. 3033.

Ignition

1 · Wells HG, *Travels of a Republican Radical in Search of Hot Water*, Harmondsworth, UK, Penguin, 1939.

2 · Davies C, *Analysis of Fire Causes on or Threatening Public Land in Victoria 1976/77–1995/96*, Fire Management Branch, Department of Natural Resources and Environment, Melbourne, 1997.

3 · Australian Bureau of Statistics, 'Environment: Bushfires', *1301.0 Year Book Australia 2004*, Australian Bureau of Statistics, Canberra, 2004.

4 · Country Fire Authority, *The Major Fires Originating 16th February, 1983*, Country Fire Authority, Melbourne, 1983.

5 · Murray R & White K, *State of Fire: A History of Volunteer Firefighting and the Country Fire Authority in Victoria*, Hargreen Publishing Company, Melbourne, 1995, p. 229.

6 · Houston C & Bachelard M, 'Fallen Powerlines May Trigger Class Action', *Sydney Morning Herald*, 15 February 2009, viewed 11 October 2009, smh.com.au.

7 · Webster J, *The Complete Australian Bushfire Book*, Penguin Books, Melbourne, 1989, p. 130.

8 · Beach J, 10 September 2009 in Teague B, McLeod R & Pascoe S, *Transcript of Proceedings, 2009 Victorian Bushfires Royal Commission*, Melbourne, CRS Wordwave Pty Ltd, 2009, p. 6810.

9 · Byrant C, *Understanding Bushfire: Trends in Deliberate Vegetation Fires in Australia*, Australian Institute of Criminology, Canberra, 2008.

10 · Willis M, 'Bushfire Arson: A Review of the Literature', Australian Institute of Criminology, Canberra, 2004.

11 · Willis M, 'Bushfire Arson: A Review of the Literature', Australian Institute of Criminology, Canberra, 2004.

Fuel

1 · Chen K & McAneney J, 'Quantifying Bushfire Penetration into Urban Areas in Australia', *Geophysical Research Letters*, vol. 31, no. L12212, 2004, pp. 1–4.

2 · Burns MR, Taylor JG & Hogan JT, 'Integrative Healing: The Importance of Community Collaboration in Postfire Recovery and

Prefire Planning', in Martin WE, Raish C & Kent B (eds), *Wildfire Risk: Human Perceptions and Management Implications,* RFF Press, Washington, DC, 2008, pp. 81–97.

3 · Cohn PJ, Williams DR & Carrol MS, 'Wildland-Urban Interface Residents' Views on Risk and Attribution', in Martin WE, Raish C & Kent B (eds), *Wildfire Risk: Human Perceptions and Management Implications*, RFF Press, Washington, DC, 2008, p. 36.

4 · Anon., 'The Black Forest of Victoria', *Australasian,* vol. 3, no. 1, 1851, pp. 456–72.

5 · Griffiths T, *Forests of Ash: An Environmental History,* Cambridge University Press, Melbourne, 2001.

6 · Clode D, 'The Mystery of the Reappearing Possums', *Continent of Curiosities: A Journey through Australian Natural History,* Cambridge University Press, Melbourne, 2006, pp. 72–84.

7 · Lindenmayer D, *Wildlife and Woodchips: Leadbeater's Possum; a Test Case for Sustainable Forestry,* University of New South Wales Press, Sydney, 1996.

8 · Bennett B, 'An Imperial, National and State Debate: The Rise and Near Fall of the Australian Forestry School 1927–1945', *Environment and History,* vol. 15, 2009, pp. 217–44.

9 · Bennett B, 'An Imperial, National and State Debate: The Rise and Near Fall of the Australian Forestry School 1927–1945', *Environment and History,* vol. 15, 2009, pp. 217–44.

10 · Florence R, 'The Ecological Basis of Forest Fire Management in New South Wales', in Attiwill PM et al. (eds), *The Burning Continent: Forest Ecosystems and Fire Management in Australia*, Perth, Institute of Public Affairs, 1994, pp. 15–33.

11 · Forests Commission, *Annual Reports, 1919–1974*, Forests Commission (Victoria), Melbourne, 1974.

12 · Young GF, *Under the Coolibah Tree,* Andrew Melrose, London, 1953, p. 234.

13 · Griffiths, *Forests of Ash,* p. 139.

14 · Pyne S, *Burning Bush: A Fire History of Australia,* University of Washington Press, Seattle, 1991, pp. 335–7.

15 · Clode D, *As If for a Thousand Years: A History of Victoria's Land Conservation and Environment Conservation Councils,* Victorian Environmental Assessment Council, Melbourne, 2006.

16 · Ward DJ & van Didden G, *Reconstructing the Fire History of the Jarrah Forest of South-western Australia: A Report to Environment Australia under the Regional Forest Agreement December 1997,* Environment Australia, Perth, 1997.

17 · Pfitzner C, *Hotspots Fire Project: Case Study: Learning and Living with Fire—The Currawinya Story,* Nature Conservation Council of New South Wales, Sydney, 2006.

18 · Simmons D, Adams R & Stoner J, 'Fuels of the Future: The Challenge of New Fuel Types', in *Bushfire Conference 2006—Life in a Fire-prone Environment: Translating Science into Practice*, Brisbane, 2006.

19 · Latz P, *The Flaming Desert: Arid Australia—A Fire-shaped Landscape*, Peter Latz, Alice Springs, 2007.

20 · Downes SJ, Handasyde KA & Elgar MA, 'The Use of Corridors by Mammals in Fragmented Australian Eucalypt Forests', *Conservation Biology*, vol. 11, no. 3, 1997, pp. 718–26.

The Bush Brigades

1 · 'Devotions: XVII Meditation' in Alford H (ed.), *The Works of John Donne: with a Memoir of His Life*, vol. III, John W. Parker, London, p. 575.

2 · Wood KA, *Ready, Aye Ready: A History of the Volunteer Fire Brigade Movement in Western Australia*, Western Australian Volunteer Fire Brigades Association, Perth, 1989.

3 · Wood KA, *Ready, Aye Ready: A History of the Volunteer Fire Brigade Movement in Western Australia*, Western Australian Volunteer Fire Brigades Association, Perth, 1989.

4 · Murray R & White K, *State of Fire: A History of Volunteer Firefighting and the Country Fire Authority in Victoria*, Hargreen Publishing Company, Melbourne, 1995, p. 33.

5 · *Brisbane Courier*, 16 February 1892, p. 5; 13 January 1892, p. 5.

6 · *West Australian*, 15 April 1897, p. 3.

7 · Hilverda S, *Hear the Siren: The First Fifty Years of the Panton Hill Volunteer Fire Brigade 1939–1990*, Panton Hill Volunteer Fire Brigade, Melbourne, 1990.

8 · Murray & White, *State of Fire*, p. 71.

9 · Hilverda, *Hear the Siren*; Motschall MS, *Wild Wood Days at Panton Hill*, Melbourne, Braidwood Press, 1984, p. 55.

10 · Sim J, *Fire and Fire Brigades*, View Productions, Sydney, 1984.

11 · Murray & White, *State of Fire*, p. 86.

12 · Pyne S, 'Black Saturday: The Sequel', *Peeling Back the Bark*, Forest History Society, 10 February 2009.

13 · Junger S, *Fire*, Fourth Estate, London, 2001.

14 · Junger S, *Fire*, Fourth Estate, London, 2001.

15 · Pyne S, *Burning Bush: A Fire History of Australia*, University of Washington Press, Seattle, 1991, pp. 356–60.

Be Prepared

1 · Country Fire Authority (Victoria), 'Total Fire Ban Declaration History', data 1945–2015, data.vic.gov.au.

2 · Sharpies JJ, McRae RHD, Weber RO & Gill AM, 'A Simple Index for Assessing Fire Danger Rating', *Environmental Modelling & Software*, vol. 24, no. 6, 2009, pp. 764–74.
3 · McAneney KJ, 'Australian Bushfires: Quantifying and Pricing the Risk to Residential Properties', paper presented at the Proceedings of the Symposium on Planning for Natural Hazards—How Can We Mitigate the Impacts? University of Wollongong, 2–5 February 2005.
4 · Daniel T, 'Managing Individual Response: Lessons from Public Health Risk Behaviour Research', in Martin WE, Raish C & Kent B (eds), *Wildfire Risk: Human Perceptions and Management Implications*, RLE Press, Washington, DC, 2008, pp. 103–16.
5 · Haynes K, Tibbits A, Coates L, Ganewatta G, Handmer J & McAneney J, *One Hundred Years of Australian Civilian Bushfire Fatalities: Exploring the Trends in Relation to the Stay or Go Policy*, Bushfire Cooperative Research Centre, Melbourne, 2008.
6 · Webster J, *The Complete Australian Bushfire Book*, Penguin Books, Melbourne, 1989; Schauble J, *The Australian Bushfire Safety Guide: The Essential Guide for Every Home*, HarperCollins, Sydney, 2004.
7 · Country Fire Authority (Victoria), *Living in the Bush*, 2nd edn, Country Fire Authority, Melbourne, 2008, p. 20.
8 · Cheney in Fahy M, *Black Tuesday: 1967 Tasmania Bushfires*, DVD, 29 mins, Bushfire CRC, Australia, 2005.
9 · Webster, *The Complete Australian Bushfire Book, p. 139.*
10 · Australian National Botanic Gardens, 'Fires, Gardens and Fire Retardant Plants—A Bibliography', 2006, viewed 27 September 2009, anbg.gov.au.
11 · Leonard JE, Blanchi R & Leicester RH, 'On the Development of a Risk-model for Bushfire Attack on Housing', *Risk Conference 2004—Melbourne*, Melbourne, 2004.
12 · Country Fire Service, *Landscaping for Fire Protection*, Community Fire Safe Wildfire Fact Sheet, No. 15, Adelaide, 2000.
13 · Fahy, *Black Tuesday.*
14 · Bazerman MH & Watkins M D, *Predictable Surprises: The Disasters You Should Have Seen Coming, and How To Prevent Them (Leadership for the Common Good)*, Harvard Business School Publishing Corporation, Boston, 2004.
15 · Spratt D & Sutton P, *Climate Code Red*, Scribe, Melbourne, 2009, p. 163.

Facing the Flames

1 · Ripley A, *The Unthinkable: Who Survives When Disaster Strikes—and Why*, Random House, London, 2008.
2 · Leach J, 'Why People "Freeze" in an Emergency: Temporal and

Cognitive Constraints on Survival Responses', *Aviation Space and Environmental Medicine*, vol. 75, no. 6, 2004, pp. 539–42.

3 · Kastenbaum R & Aisenberg R, *The Psychology of Death*, Duckworth, London, 1974, p. 381.

4 · Kumagai Y & Nojima Y, 'Urbanization and Disaster Mitigation in Tokyo', *Crucibles of Hazard: Mega-cities and Disasters in Transition*, United Nations University Press, Tokyo, 1999, pp. 56–91.

5 · Ripley, *The Unthinkable.*

6 · Bracha HS, 'Human Brain Evolution and the "Neuroevolutionary Time-depth Principle": Implications of the Reclassification of Fear-Circuitry-Related Traits in DSM-V and for Studying Resilience to Warzone-Related Posttraumatic Stress Disorder', *Progress in Neuro-Psychopharmacology and Biological Psychiatry*, vol. 30, 2006, pp. 827–53.

7 · Bracha HS, 'Human Brain Evolution and the "Neuroevolutionary Time-depth Principle": Implications of the Reclassification of Fear-Circuitry-Related Traits in DSM-V and for Studying Resilience to Warzone-Related Posttraumatic Stress Disorder', *Progress in Neuro-Psychopharmacology and Biological Psychiatry*, vol. 30, 2006, pp. 827–53.

8 · Haynes K, Tibbits A, Coates L, Ganewatta G, Handmer J & McAneney J, *One Hundred Years of Australian Civilian Bushfire Fatalities: Exploring the Trends in Relation to the Stay or Go Policy*, Bushfire Cooperative Research Centre, Melbourne, 2008.

9 · Kenney WL, DeGroot DW & Holowatz LA, 'Extremes of Human Heat Tolerance: Life at the Precipice of Thermoregulatory Failure', *Journal of Thermal Biology*, vol. 29, no. 7–8, 2004, pp. 479–85.

10 · Noble WS, *Ordeal by Fire: The Week a State Burned Up*, Jenkin Buxton Printers, Melbourne, 1977, p. 25.

11 · Grillon C, Pellowski M, Merikangas KR & Davis M, 'Darkness Facilitates the Acoustic Startle Reflex in Humans', *Biological Psychiatry*, vol. 42, no. 6, 1997, pp. 453–60.

12 · McCaffrey S, 'Understanding Public Perspectives of Wildfire Risk', in Martin WE, Raish C & Kent B (eds), *Wildfire Risk: Human Perceptions and Management Implications*, RFF Press, Washington, DC, 2008, pp.11–22.

13 · Murphy SA, Beaton RD, Cain K & Pike K, 'Gender Differences in Fire Fighter Job Stressors and Symptoms and Stress', *Women and Health*, vol. 22, no. 2, 1994, pp. 55–69.

14 · Fowlkes MR & Miller PY, 'Chemical and Community at Love Canal', in Johnson BB & Covello VT (eds), *The Social and Cultural Construction of Risk*, D. Reidel Publishing Company, Dordrecht, 1987.

15 · Mannix I, *Great Australian Bushfire Stories*, ABC Books, Sydney, 2008, p. 166.

16 · Krusel N & Petris SN, *A Study of Civilian Deaths in the 1983 Ash*

Wednesday Bushfires Victoria, Australia, Country Fire Authority, Melbourne, 1992.

17 · Haynes, Tibbits, Coates, Ganewatta, Handmer & McAneney, *One Hundred Years of Australian Civilian Bushfire Fatalities.*

18 · Cheney P, Gould J & McCaw L, 'The Dead-man Zone—A Neglected Area of Firefighter Safety', *Australian Forestry*, vol. 64, no. 1, 2001, pp. 45–50.

Leaving Late

1 · Leonard JE, Blanchi R & Leicester RH, 'On the Development of a Risk-model for Bushfire Attack on Housing', *Risk Conference 2004—Melbourne*, Melbourne, 2004.

2 · Haynes K, Tibbits A, Coates L, Ganewatta G, Handmer J & McAneney J, *One Hundred Years of Australian Civilian Bushfire Fatalities: Exploring the Trends in Relation to the Stay or Go policy*, Bushfire Cooperative Research Centre, Melbourne, 2008.

3 · Haynes K, Tibbits A, Coates L, Ganewatta G, Handmer J & McAneney J, *One Hundred Years of Australian Civilian Bushfire Fatalities: Exploring the Trends in Relation to the Stay or Go policy*, Bushfire Cooperative Research Centre, Melbourne, 2008.

4 · Haynes K, Tibbits A, Coates L, Ganewatta G, Handmer J & McAneney J, *One Hundred Years of Australian Civilian Bushfire Fatalities: Exploring the Trends in Relation to the Stay or Go policy*, Bushfire Cooperative Research Centre, Melbourne, 2008.

5 · Bushfire CRC, 'Guidance for People in Vehicles During Bushfires', *Fire Note*, no. 8, 2006, p. 6.

6 · Sorenson JH, Shumpert BL & Vogt BM, 'Planning for Protective Action Decision Making: Evacuate or Shelter-in-Place', *Journal of Hazardous Materials*, vol. A109, 2004, pp. 1–11.

7 · Sekizawa A, 'Human Factors in Fire Safety Measures', *Fire Science and Technology*, vol. 23, no. 4, 2004, pp. 292–302.

8 · Alsnih R & Stopher PR, *Review of Procedures Associated with Devising Emergency Evacuation Plans*, Transportation Research Board National Research Council, Washington, DC, 2004; Cova TJ & Johnson JP, 'Microsimulation of Neighborhood Evacuations in the Urban-Wildland Interface', *Environment and Planning A*, vol. 34, no. 12, 2002, pp. 2211–29.

9 · Cohn PJ, Carroll MS & Kumagai Y, 'Evacuation Behavior During Wildfires: Results of Three Case Studies', *Western Journal of Applied Forestry*, vol. 21, no. 1, 2006, pp. 39–48.

10 · Dennison PE, Cova TJ & Mortiz MA, 'WUIVAC: A Wildland-Urban Interface Evacuation Trigger Model Applied in Strategic Wildfire Scenarios', *Natural Hazards*, vol. 41, no. 1, 2007, pp. 181–99.

Shelter

1 · Schapel AE, *Inquest into the Deaths of Star Ellen Borlase, Jack Morley Borlase, Helen Kald Castle, Judith Maud Griffith, Jody Maria Kay, Graham Joseph Russell, Zoe Russell-Kay, Trent Alan Murnane and Neil George Richardson,* Coroner's Court of South Australia, Adelaide, 2007.

2 · Cowan J, 'Many Bushfire Victims Died in Their Bathrooms: Report' *ABC News*, 26 June 2009, viewed 29 September 2009, abc.net.au.

3 · Stretton L, *Report of the Royal Commission to Inquire into the Causes of and Measures Taken to Prevent the Bush Fires of January, 1939, and to Protect Life and Property and the Measures to be Taken to Prevent Bush Fires in Victoria and to Protect Life and Property in the Event of Future Bush Fires,* State Government of Victoria, Melbourne, 1939.

4 · Stretton L, *Report of the Royal Commission to Inquire into the Causes of and Measures Taken to Prevent the Bush Fires of January, 1939, and to Protect Life and Property and the Measures to be Taken to Prevent Bush Fires in Victoria and to Protect Life and Property in the Event of Future Bush Fires,* State Government of Victoria, Melbourne, 1939.

5 · Australian Building Codes Board, 'National Standards for Bushfire Bunkers', media release, 24 September 2009.

6 · Teague B, McLeod R & Pascoe S, *Transcript of Proceedings, 2009 Victorian Bushfires Royal Commission,* CRS Wordwave Pty Ltd, Melbourne, 24 June 2009, pp. 3993–5.

7 · Webster J, *The Complete Australian Bushfire Book,* Penguin Books, Melbourne, 1989.

8 · Quoted in *Canberra Times*, 23 January 1939, p. 2.

The Need to Know

1 · Abernathy J, *Inquests into the Deaths of John William Dean, Michael Bannister, James Michael Lawler, Glyn Roderick Charles, Phillip Raymond Charles Skeggs, Bruce Raymond Guy during the 1998 Sydney to Hobart Yacht Race*, Sydney, State Coroner, 2000.

2 · Mileti DS & Peek L, 'The Social Psychology of Public Response to Warnings of a Nuclear Power Plant Accident', *Journal of Hazardous Materials*, vol. 75, no. 2–3, 2000, pp. 181–94.

3 · Mileti DS, Bandy R, Bourque LB, Johnson A, Kano M, Peek L, Sutton J & Wood M, *Annotated Bibliography for Public Risk Communication on Warnings for Public Protective Action Response and Public Education,* Boulder Natural Hazards Centre, University of Colorado, Boulder, 2006.

4 · Mileti & Peek, 'The Social Psychology of Public Response'.

5 · Indian J, 'The Concept of Local Knowledge in Rural Australian Fire Management', in Handmer J & Haynes K (eds), *Community Bushfire Safety,* CSIRO Publishing, Melbourne, 2008, pp. 35–45.

6 · Kapuchi N & van Wart M, 'Making Matters Worse: An Anatomy of Leadership Failures in Managing Catastrophic Events', *Administration and Society,* vol. 40, no. 7, 2008, pp. 711–40.

Recovery and Resilience

1 · Melville H, *Moby Dick,* Spark Educational Publishing, New York, 2003, p. 492.
2 · Gordon R, 'Thirty Years of Trauma Work: Clarifying and Broadening the Consequences of Trauma', *Psychotherapy in Australia,* vol. 13, no. 3, 2007, pp. 12–19.
3 · Byrne ME, Lerias D & Sullivan NL, 'Predicting Vicarious Traumatization in Those Indirectly Exposed to Bushfires', *Stress and Health,* vol. 22, 2006, pp. 167–77.
4 · Burns MR, Taylor JG & Hogan JT, 'Integrative Healing: The Importance of Community Collaboration in Postfire Recovery and Prefire Planning', in Martin WE, Raish C & Kent B (eds), *Wildfire Risk: Human Perceptions and Management Implications,* RFF Press, Washington, DC, 2008, p. 88.
5 · Burns, Taylor & Hogan, 'Integrative Healing'.
6 · Gordon, 'Thirty Years of Trauma Work'.
7 · Gordon R, 'Community Process and the Recovery Environment Following Emergency', *Environmental Health,* vol. 4, no. 1, 2004, pp. 9–24.
8 · Aptekar L, 'The Psychology of Disaster Victims', *Environmental Disasters in Global Perspective,* Macmillan, New York, 1994.
9 · Chamberlain ER, Doube L, Milne G, Rolls M & Western JS, *The Experience of Cyclone Tracy,* Australian Government Publishing Service, Canberra, 1981.
10 · Carroll MS, Cohn PJ, Seesholtz DN & Higgins LL, 'Fire as a Galvanising and Fragmenting Influence on Communities: The Case of the Rodeo–Chediski Fire', *Society and Natural Resources,* vol. 18, no. 4, 2005, pp. 301–20.
11 · Gordon, 'Community Process and the Recovery Environment Following Emergency'.
12 · Mannix I, *Great Australian Bushfire Stories,* ABC Books, Sydney, 2008, pp. 44–5.

What Next?

1 · Fendokin MA, Gehling JG, Grey K, Narbonne GM & Vickers-Rich P, *The Rise of Animals: Evolution and Diversification of the Kingdom Animalia,* Johns Hopkins University Press, Baltimore, 2009.
2 · White ME, *After the Greening,* Kangaroo Press, Sydney, 1994.
3 · Hennessy K, Lucas C, Nicholls N, Bothols J, Suppiah R & Ricketts J,

Climate Change Impacts of Fire-weather in South-east Australia, CSIRO Marine and Atmospheric Research, Melbourne, 2005.

4 · Hennessy K, Lucas C, Nicholls N, Bothols J, Suppiah R & Ricketts J, *Climate Change Impacts of Fire-weather in South-east Australia*, CSIRO Marine and Atmospheric Research, Melbourne, 2005.

5 · Cannon M, *Melbourne after the Gold Rush*, Loch Haven, Melbourne, 1993.

6 · Dwyer G & Esnouf G, 'Inquiries and Reviews in Victoria: Key Recommendations and Outcomes 1939–2008', in *International Bushfire Research Conference and 15th Annual AFAC Conference*, Adelaide Convention Centre, 2008.

7 · Daniel T, 'Managing Individual Response: Lessons from Public Health Risk Behaviour Research', in Martin WE, Raish C & Kent B (eds), *Wildfire Risk: Human Perceptions and Management Implications*, RFF Press, Washington, DC, 2008, pp. 103–16.

8 · Bushfire CRC, *Victorian 2009 Bushfire Research Response Interim Report*, Bushfire CRC, Melbourne, 2009.

PRAISE FOR
A FUTURE IN FLAMES

'Bushfires have always threatened and always will. *A Future in Flames* makes sense of it all.'

—Christine Nixon

'Concise and insightful ... a thought-provoking analysis.'

—Kevin O'Loughlin

'If ever there was a book for its time, it is *A Future in Flames* ... [a] superb volume.'

—Adrian Hyland, *The Age*

'A thoughtful and broad-ranging work.'

—Roy Williams, *The Australian*

'A discomforting but hugely important book that deserves to be widely read.'

—*The Courier Mail*

'The beauty of this book is that it provides sensible information, and blames nobody.'

—Lucy Sussex, *The Sunday Age*

'Written with an assured style that moves from lyrical to urgent.'

—Caroline Baum, *Better Homes and Gardens*

COPYRIGHT

First published in 2010 by Melbourne University Press

This edition published in 2018
by Ligature Pty Limited
34 Campbell St · Balmain NSW 2041
www.ligatu.re · mail@ligatu.re

e-book ISBN: 9780648140740

ligature *un*tapped

This print edition published in collaboration with Brio Books in, an imprint of Booktopia Group Ltd

Level 6, 1A Homebush Bay Drive · Rhodes NSW 2138 · Australia

Print ISBN: 9781761280894

briobooks.com.au

b brio BOOKS

MIX
Paper from responsible sources
FSC
www.fsc.org
FSC® C008194

The paper in this book is FSC® certified. FSC® promotes environmentally responsible, socially beneficial and economically viable management of the world's forests.